Sterilization and Preservation

Mamata Mukhopadhayay ·
Anuradha Chatterjee

Sterilization and Preservation

Applications of Supercritical Carbon Dioxide

Ane Books Pvt. Ltd.

Springer

Dr. Mamata Mukhopadhayay
Emeritus Professor
Department of Chemical Engineering
Indian Institute of Technology
Bombay, India

Dr. Anuradha Chatterjee
Former Research Associate
Department of Chemical Engineering
Indian Institute of Technology
Bombay, India

ISBN 978-3-031-17372-1 ISBN 978-3-031-17370-7 (eBook)
https://doi.org/10.1007/978-3-031-17370-7

Jointly published with Ane Books Pvt. Ltd.
In addition to this printed edition, there is a local printed edition of this work available via Ane Books in South Asia (India, Pakistan, Sri Lanka, Bangladesh, Nepal and Bhutan) and Africa (all countries in the African subcontinent).
ISBN of the Co-Publisher's edition: 978-93-90658-49-7

This Springer imprint is published by the registered company Springer Nature Switzerland AG
The registered company address is: Gewerbestrasse 11, 6330 Cham, Switzerland

Dedication

This book is dedicated to those Covid-19 warriors, selfless medical doctors, and health workers who fought relentlessly and sacrificed their lives to save millions, due to the unprecedented pandemic.

Preface

The pandemic that started in 2020, had the entire world in its clutches. Mankind had not tackled a pandemic of this proportion in the last 102years; to make matters worse, the entire medical fraternity and pharma industry were reeling under the stress of this unknown terror, for they knew not how to manage this unknown enemy. The motivation for writing this book arises from these unprecedented times of the pandemic that brought with it chaos and crisis creating an environment of scare and scarcity, and which brought us to the fore of a few issues that need urgent attention of scientific researchers and engineers all over the world. The most crucial sectors of concern include foods and pharmaceuticals needed for human survival, biomaterials and medical devices essential for protection of health and lives, and biomedical waste management for safety of the people and environment. Recently the food, pharmaceutical and medical fraternity have been making all out efforts globally for their availability, sustainability, safety, conservation, transportation, and distribution, and have been urgently looking for an effective sterilization and preservation technology to meet these challenges in an economically viable way.

Food products and pharmaceuticals are functional and effective only if the nutritive and bioactive ingredients remain active, and these bioactive molecules are usually sensitive to heat and harsh treatment procedures. Also, in recent years pharma and medical industries have started relying heavily on biomaterials, which are complex in morphology and composition, and are highly sensitive to processing conditions. Thus, the classical method of sterilization is slowly but surely becoming redundant, paving the path to explore more benign technologies on an urgent basis. Additionally, biomedical waste management poses a huge concern globally, especially during the present times when COVID-19 has paralysed the world. India alone has generated almost 18,000 tonnes of COVID-19 biomedical wastes in last four months (as on October 2020), and it is estimated by the UN that each year around 5.2 million people, including 4 million children, would die owing to diseases related to improper disposal of biomedical wastes, especially when the medical wastes have risen almost ten times and pose a huge threat in terms of cross contamination. This alone emphasises the importance to make the biomedical waste benign, recyclable, and non-spreader of disease. The newly emerging supercritical carbon dioxide (scCO_2) technology is the most desirable option to tackle these issues. Accordingly potential applications of this technology will be presented in this book, in order to address the problems of sterilization and preservation of foods, biomaterials, and medical wastes management to overcome the burden of related diseases and death, without having any negative impact on the environment.

This book will discuss in detail about the insight on the scCO_2 technology, its advantages over the prevalent methods for sterilization and preservation, the processing techniques and selection of process parameters, and the effectiveness of the use of this technology for the aforementioned objectives, citing a few case studies. The proposed book would cater to a large number of researchers in academic institutions, research organisations, and Food and Pharma Industries. It could also be utilized as a textbook for students aspiring for specialised courses in the disciplines of Food Processing and Preservation, Sterilization of Biomaterials and Biomedical Devices, and Management of Biological and Biomedical Wastes.

The authors wish to acknowledge with gratitude the constant inspiration and technical assistance received from Dr. Nripendu Dutta, the younger brother of the first author, the library assistance provided by Dr. Neha Singh of IIT Bombay, and constant encouragement and cooperation obtained from Master Arko Chatterjee and Mr. Indrayan Chatterjee, the son and husband respectively, of the second author towards completion of the book.

Mamata Mukhopadhyay Anuradha Chatterjee

Contents

Dedication v

Preface vii

1. Introduction to Sterilization and Preservation Using Supercritical CO_2 1

1.1 Relevance of Sterilization and Preservation of Food Products 1

1.1.1 Scope of Sterilization and Preservation 2

1.1.2 Need for a Green Effective Technology in Food Industries 3

1.2 Relevance of Sterilization in Medical Industries 4

1.2.1 Emergence of New Biomaterials 4

1.2.2 Need for a Green Effective Technology for Biomaterials and Medical Devices 4

1.3 Relevance of Sterilization with $ScCO_2$ for Management of Clinical Solid Wastes 5

References 6

2. Classification of Foods, Biomaterials and Microorganisms 9

2.1 Food Constituents 9

2.1.1 Water 9

2.1.2 Minerals 10

2.1.3 Carbohydrates 11

2.1.4 Fats and Lipids 12

2.1.5 Organic Acids 12

2.1.6 Nitrogen Containing Substances 13

2.1.7 Vitamins 13
2.1.8 Proteins 13
2.1.9 Enzymes 13
2.2 Classification of Foods by Characteristic Properties 19
2.2.1 Texture 19
2.2.2 Color 22
2.2.3 Flavor 22
2.2.4 Nutritional Aspects 22
2.2.5 Food Extracts and Phyto-Pharmaceuticals 22
2.2.6 Safety, Diversity and Complexity 23
2.3 Factors Causing Degradation and Spoilage of Foods 24
2.3.1 Physical Factors 24
2.3.2 Chemical Factors 25
2.3.3 Microorganisms 25
2.4 Undesirable Enzymes and Hurdles in Food Industry 27
2.5 Classification of Biomaterials by Their Activity 27
2.5.1 Natural Polymers 28
2.5.2 Artificial Polymers 30
2.5.3 Medical Implants 31
2.6 Microorganisms Involved in Contamination of Biomaterials 32
2.6.1 Sources of Contamination 33
2.6.2 Classification of Agents of Contamination 34
2.7 Contamination from Clinical Waste and Microflora 35
2.7.1 Contaminants in Clinical Solid Waste 36
2.7.2 Contaminants in Liquid Medical Waste 37
References 40

3. Characterization Methods and Evaluation of Sterility 43
3.1 Relevance of Microbial Analysis in Food Industry 43
3.1.1 Sources of Contamination 44
3.1.2 Standard Testing Methods for Microbes 44
3.1.3 Environment Scanning Electron Microscope (ESEM) 50

3.1.4 Limits of Detection and Requirements of Inactivation 51
3.1.5 Standard Testing Methods for Enzyme Activity 51
3.2 Relevance of Microbial Analysis in Biomaterials 54
3.2.1 Factors for Contamination 54
3.2.2 Detection and Testing Methods 54
3.2.3 Standards of Sterility and Sterilization Efficiency 56
3.3 Standards of Sterility Levels for Clinical Waste Management 58
References 58

4. Conventional Processes for Sterilization and Preservation 61
4.1 Food Preservation 61
4.2 Food Preservation Techniques 62
4.2.1 Reduction in Temperature 63
4.2.2 Reduction in pH 64
4.2.3 Reduction in Oxygen Level in Modified Atmosphere 65
4.2.4 Reduction of Water Activity 66
4.2.5 Application of Heat 68
4.2.6 Irradiation with Radionucleotide 71
4.2.7 Pulsed Electric Field and High Voltage Arch Discharge 71
4.2.8 Ultra High Pressure and Ultra Sound 72
4.2.9 Plasma 72
4.2.10 Oscillating Magnetic field 73
4.3 Sterilization of Biomaterials and Medical Devices 74
4.3.1 Thermal Sterilization 75
4.3.2 Ethylene Oxide (ETO) Gas Sterilization 75
4.3.3 Vaporized Hydrogen Peroxide (VHP) Sterilization. 76
4.3.4 Hydrogen Peroxide Gas Plasma Sterilization 77
4.3.5 Peracetic Acid Sterilization 77
4.3.6 Ozone Immersion Sterilization 78
4.3.7 High Energy-Gamma Irradiation. 78
4.3.8 Electron Beam (E-Beam) Sterilization: 79
4.3.9 Microwave Sterilization 79
4.4 Safety and Effectiveness of the Processes 79
References 82

5. Processing with Supercritical Carbon Dioxide **85**
5.1 Unique Advantages of $ScCO_2$-based-Technologies 85
5.2 Characteristics of Supercritical Carbon Dioxide 89
5.2.1 Path for Attaining Supercritical State 89
5.2.2 Unique Advantages and Properties as a Solvent 90
5.3 Supercritical Fluid Extraction Process 91
5.4 Commercial Scale Supercritical Fluid Extraction Process 94
5.5 Commercial Applications of SCF Extraction from Natural Products 97
5.6 Other $ScCO_2$- based Processes 98
5.6.1 Micronisation, Microencapsulation, and Impregnation for Drug Delivery 99
5.6.2 Synthesis of Novel Materials, Specialty Chemicals and Polymers in $ScCO_{2}$ Medium 102
5.6.3 Biochemical Reactions in $ScCO_2$ 103
5.7 Conclusion 103
References 104

6. Sterilization and Preservation of Solid Foods with Supercritical CO_2 **107**
6.1 Performance Evaluation of $ScCO_2$ for Sterilization and Stabilization 108
6.1.1 Sterilization and Stabilization of Rice Bran 108
6.1.2 Sterilization and Stabilization of Soybeans 114
6.2 Sterilization of Wheat 121
6.3 Sterilization of Coconut 122
6.4 Sterilization of Meats 122
References: 124

7. Sterilization and Preservation of Liquid Foods with Supercritical CO_2 **127**
7.1 Tomato Puree 128
7.1.1 Characteristics of Tomato Juice 128
7.1.2 Conventional Preservation Processes 130
7.1.3 Preservation by Treatment with $ScCO_2$ 130
7.2 Sugarcane Juice 134
7.2.1 Characteristics of Sugarcane Juice 134

7.2.2 Conventional Preservation Methods 136
7.2.3 Preservation and Stabilization with $ScCO_2$ 136
7.3 Aloe Vera Juice 137
7.3.1 Characteristics of Aloe Vera Juice 138
7.3.2 Preservation of Aloe Vera Juice 140
7.3.3 Preservation and Stabilization of Aloe Vera Juice with $ScCO_2$ 141
7.4 Cow Milk 144
7.4.1 Characteristics of Cow Milk 146
7.4.2 Preservation and Pasteurization of Milk 147
7.4.3 Preservation of Cow Milk by $ScCO_2$ Treatment 148
7.5 Apple Juice 150
7.5.1 Characteristics 150
7.5.2 Conventional Preservation Process 151
7.5.3 Preservation with $ScCO_2$ Treatment 152
References 153

8. Mechanism of Sterilization and Preservation Using Supercritical CO_2 157
8.1 Bactericidal Effect of CO_2 157
8.2 Mechanism for Inactivation of Vegetative Microorganisms 158
8.2.1 Cell Wall Rupture 159
8.2.2 Physiological Inactivation 160
8.2.3 Biological Stress 164
8.3 Inactivation Mechanism of Bacterial Spores 165
8.4 Mechanism of Virus Inactivation 166
8.5 Mechanism of Inactivation of Enzymes 166
8.5.1 Inhibitory Effect of Molecular CO_2 166
8.5.2 Effect of Characteristics of $ScCO_2$ 167
8.5.3 Effect of Pressurization-depressurization cycling 167
References 168

9. Sterilization of Biomaterials and Medical Devices with Supercritical CO_2 171
9.1 Safety and Success of Medical Devices 171
9.2 Methods of Contamination and Infectious Agents 172
9.3 Factors Involved in Sterilization of Biomaterials 173

9.4 Current Methods of Sterilization and Disinfection 174
9.5 Terminal Sterilization Using $ScCO_2$ 177
9.6 Effectiveness of $ScCO_2$ Treatment of Biomaterial with Select Additives 178
9.7 Inactivation of Virus 179
9.8 Sterilization Process for Rapid Inactivation of Bacterial Endospores 180
9.9 Elimination of Endotoxins and Pyrogens 182
9.10 Sterilization of Biologically Active Large Molecules and Implant Materials 183
References 185

10. Clinical Sold Waste Management with $ScCO_2$ Sterilization Technology 187
10.1 Generation, Handling and Safe Disposal of Clinical Solid Wastes 187
10.2 Recycle-Reuse of CSW after Sterilization 189
10.3 Pathogenic Microorganisms Present in Clinical Wastes 190
10.3.1 Pathogenic Bacteria in Clinical Wastes 191
10.3.2 Infectious Fungi in Clinical Wastes 192
10.4 Thermal methods for Inactivation and Neutralization 193
10.4.1 Incineration 193
10.4.2 Steam Sterilization 194
10.5 Nonthermal Sterilization Technology Using $ScCO_2$ 195
10.5.1 Inactivation of Micro-organisms Using $ScCO_2$ 195
10.5.2 Inactivation of Fungi Using $ScCO_2$ 197
10.6 Sterilization-Reuse of Personal Protective Equipment (PPE) Using $ScCO_2$ Technology 199
10.7 Environment and Health Protection Using $ScCO_2$ Sterilization Technology 201
References 203

1

Introduction to Sterilization and Preservation Using Supercritical CO_2

This book covers sterilization and preservation of three broad categories of substances, namely, (i) foods and pharmaceuticals, (ii) biomaterials and medical devices, and (iii) clinical solid wastes. Each of these industries are presently playing distinctive roles in contributing to the immense pressure of present-day challenges, especially during this pandemic situation when the race is against time to save lives as well as environment. This compilation gives an insight to the processing technology using supercritical carbon dioxide ($scCO_2$) for effective sterilization and preservation as a hall mark for adopting a single process based on $scCO_2$. It also highlights the relevance of this green technology to alleviate the recent problems and draws attention of scientists and engineers, researchers and policy makers, and administrative authorities, justifying its early implementation.

1.1 Relevance of Sterilization and Preservation of Food Products

Food is a basic necessity for survival of any living entity. Food supply and management eventually control a country's economic status and political developments. India is the second most populated country in the world growing at the annual rate of 1.2% with 17.7% (as estimated in April 2021) of the world's population. India is projected to surpass China to become the world's most populous country by 2024. The present population in India is estimated to have crossed 1.39 billion in April 2021. It is expected to become the first country to be home to more than 1.5 billion people by 2030.

M. Mukhopadhayay and A. Chatterjee, *Sterilization and Preservation*,
https://doi.org/10.1007/978-3-031-17370-7_1

The problem facing India today is not one of shortage of food but finding ways and means of managing the accumulated surplus. The surplus produced can be exported which will provide the needed economic boost to the Indian Food Processing Industry. The global market for food processing and handling equipment is expected to grow at a Compound Annual Growth Rate (CAGR) of 6.2%, making it a USD 196.6 billion industry by 2025 *[Regional Global Forecast, 2019]*. Food needs to be preserved as it begins to spoil soon after it is harvested or processed. Annually the food industry has been facing a financial loss to the tune of about 750 billion dollars, a large part of which is contributed by ineffective methods of food preservation leading to inadequate shelf life, degradation of quality, and development of rancidity. In order to be able to export, the wastage of the surplus has to be minimized and accordingly an effective preservation process has to be employed. Also, to make sure food reaches the needy and poor, it is essential that food has a better shelf life to enable its transportation to every corner of the country, where it is needed.

1.1.1 Scope of Sterilization and Preservation

Food processing implies conversion of raw materials or ingredients into consumer food products. Food processing and food preservation are very closely related terms, and often the primary or secondary objective of food processing is to achieve its preservation. A process that combines sterilization and stabilization is a quality preservation process which leads to extension of shelf life and makes the product shelf-stable. Sterilization is the process of making the substance free of live bacteria or other microorganisms without losing its essence. Stabilization is a process that allows the product to be handled under atmospheric conditions without degrading the quality. Thus for a process of sterilization and preservation it is important that it adheres to the following criteria:

i) It inactivates the microbial contaminants and makes the food safe for consumption.
ii) It inhibits the activity of undesirable enzymes that may cause degradation of the food and retains its fresh appeal.
iii) It renders the food shelf-stable and makes it transportation friendly.
iv) It eliminates any residual contaminants in the food after processing.
v) It enhances the shelf life of the food.

While achieving the above targets, the nutritional aspects, the type of micro flora, and endogenous/exogenous enzymes present in the food should be taken into consideration. Moreover, for a process to be successfully implemented in an industry, it has to be economically viable to ensure sustainability of the technology.

1.1.2 Need for a Green Effective Technology in Food Industries

Food directly concerns the health and nutrition of the people; hence it must be safe and beneficial, such that its consumption should not cause any disease. It is only imperative that a modern effective eco-friendly technology is employed in the food and drink industry to assure safety of food, with a view to countering this issue globally. However, most of the existing technologies compromise the nutritional value of the food while achieving sterilization and preservation of food. In recent times a combination of technologies is being used to minimize the loss of nutritional value, organoleptic characteristics, and color of the food to make it more appealing to consumers. Sterilization and preservation without having a negative impact on the quality and properties of the food, therefore, remain a challenge for food researchers to alleviate this problem. All these aspects entail selection of a food processing technology for effectively achieving the targeted objectives in a single process in order to make it commercially viable.

The current book presents the supercritical carbon dioxide (scCO_2) technology as one of the emerging green technologies, which has great potential in the foods and medical industries. The scCO_2 technology is benign, eco-friendly, and effective and it is one of the most promising technologies that is being widely researched upon. The unique transport properties and the GRAS status of scCO_2 forge it as an ideal choice for application in the food and drink industry. Moreover, slight modifications in its process conditions impart it the ability to selectively extract, inactivate or inhibit the component of interest. A novel sterilization, stabilization and preservation process based on the scCO_2 technology preserves both liquid and solid food products, with a few variations in the controlling parameters. The inactivation of the microorganisms and inhibition of enzymes are attributed to the physical, chemical, and biological stresses created in the environment. The process is designed to use mild operating conditions for treatment in order to retain the nutritional value, color and organoleptic characteristics of the foods. The process has been successfully demonstrated and established for a wide range of liquid and solid food products with varying pH, moisture level, and nutritional characteristics *[Mukhopadhyay and Chakraborty, 2004; 2005]*. The mechanisms of actions of scCO_2 in sterilization and preservation based on this novel benign scCO_2 technology, are unveiled in this book. A few case studies are cited *[Chakraborty, 2006]* with a view to adapting it to a wide range of food products and pharmaceuticals with increasing complexity in texture and constituents and the high sensitivity to the processing environment.

1.2 Relevance of Sterilization in Medical Industries

Over the last few decades, the use of biomaterials and medical devices in medical industry has increased manifolds. This has led to an unforeseen issue - the development of new families of diseases, which are directly related to the use of the biomaterials and medical devices. Infection related to biomaterials is one of the most peculiar and economically draining side effects of using biomaterials. As per medical survey reports, in 2010, two million people acquired nosocomial infections while in hospitals in the USA and nearly 100,000 people lost their lives owing to this acquired infection. Adoption of an eco-friendly green technology by the medical industry is imperative for effective sterilization and preservation of biomaterials.

1.2.1 Emergence of New Biomaterials

In the last few decades medical industry has made rapid progress coupled with considerable number of ground breaking research findings. Over the years there has been a substantial paradigm shift with the medical industry moving towards finding more tailor-made solutions for patients. Medical industry has diversified to become an amalgamation of biological sciences and engineering. Surgical replacements, implants, and repairs have become a commonality to address trauma, diseases, and degeneration. Knee replacements, hip joint replacements, pacemakers, silicone implants, dental implants, stents, contact lens, etc. are all examples of biomaterials that are used extensively making them almost an integral part of human lives. The earliest recorded use of biomaterials dates back to antiquity, the ancient Egyptian era when sutures were used which were made from animal sinew. The evolution of biomaterials has come a long way. The present day biomaterials are information rich and usually incorporate bioactive components, contributing heavily towards lowering mortality and improving the quality of life by providing the required support as per the disease indication and therapeutic area.

1.2.2 Need for a Green Effective Technology for Biomaterials and Medical Devices

Sterilization of biomaterials often requires processing conditions which exclude the use of classical methods involving aggressive action on the material (e.g., high temperature, strong chemicals or irradiation). As a consequence, new techniques are emerging based on different approaches to sterilization focusing on the direction of green effective processing technology. Processing with supercritical carbon dioxide ($scCO_2$) offers unique advantages that can satisfy the requirements for the preparation of biocompatible materials at affordable cost of production of medical devices. For example, the possibility of processing with $scCO_2$ at low temperatures opens up avenues to work with biopolymers and thermally sensitive components, such as, collagen and gels in the form of alginate, gellan gum, and chitosan. In addition, low surface tension, good solvency, and excellent transport properties facilitate penetration of $scCO_2$ into solid, colloidal, and microbial cell structures, enabling efficient sterilization of medical devices with the preservation of the structure and physicochemical properties. Moreover, the benign nature of CO_2 eliminates the need for post-processing and purification of the processed medical devices. This renders processing with $scCO_2$ to be considered as a versatile approach to the solution to the urgent requirement of sterilization of biomaterials and medical devices *[Veryasova et al., 2019]*.

1.3 Relevance of Sterilization with $ScCO_2$ for Management of Clinical Solid Wastes

Clinical solid wastes include any waste generated during the diagnosis, treatment or immunization of human beings or animals in hospitals and healthcare facilities, research labs or in the production or testing of biological samples. The management of clinical solid waste (CSW) poses a serious problem to the health and safety of the people mostly in densely populated nations and developing countries due to lack of strict international legislations. Inappropriate disposal methods and inadequate strategies followed during handling and disposal of CSW are causing significant health hazards and environmental pollution due to the infectious nature of the wastes. Owing to the financial constraints, most of the healthcare facilities in the developing nations are focusing on cost effective disposal methods of clinical waste and are in the process of implementation of the recycle-reuse program of CSW materials after initial sterilization at the point of their generation. Moreover, CSW disposed without being sterilized or disinfected, poses a huge threat of cross infections leading to an epidemic.

Adoption of the supercritical carbon dioxide ($scCO_2$) sterilization technology is emphasized as the priority to inactivate the infectious micro-organisms in CSW *[Efaq et al., 2015]*. The recycling-reuse program using the $scCO_2$ sterilization technology can be carried out successfully with the non-specialized clinical staffs at the point of collection. Consequently, it can reduce exposure to infectious waste, decrease labor, lower costs, and yield better compliance with regulatory bodies. Thus healthcare facilities can ensure safe collection, transportation, and disposal of biomedical wastes. At the same time they can both save money and provide a safe environment for patients, healthcare personnel, and clinical staffs.

Figure 1.1, depicts the applicability and versatility of the green technology – $scCO_2$ technology, in each of the aforementioned industries.

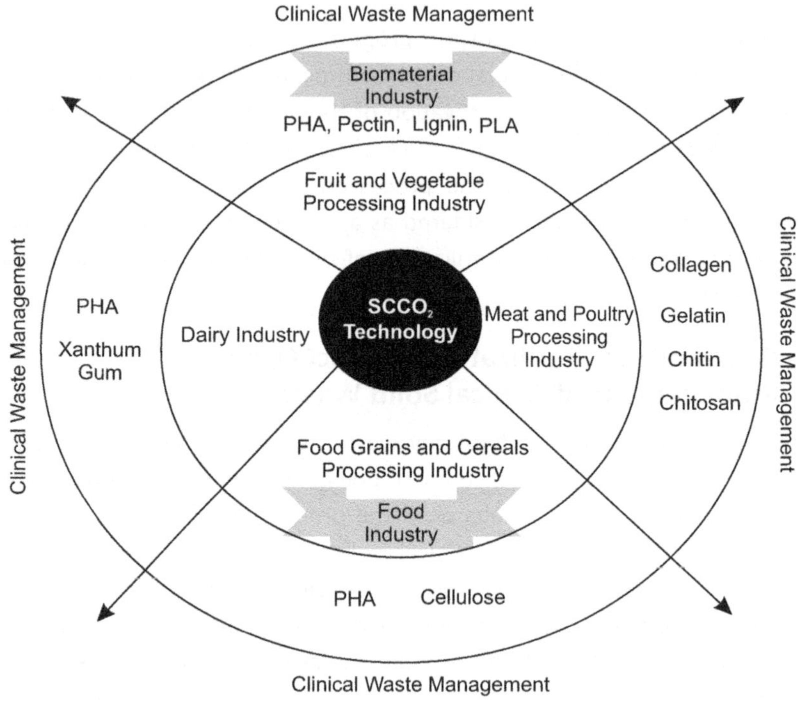

Fig. 1.1: Application areas of $scCO_2$ technology across industries.

References

- Chakraborty Anuradha, *Sterilization and Stabilization of Food Products Using Supercritical Carbon Dioxide,* Ph.D. Dissertation, Indian Institute of Technology Bombay, 2006.
- Efaq A. N, Rahman N. N. N. A, Nagao H, Al-Gheethi A. A, Shahadat M, Kadir M. O. A., *"Supercritical Carbon Dioxide as Non-Thermal Alternative Technology for Safe Handling of Clinical Wastes",* Environ. Process. 2,797–822, 2015.
- Mukhopadhyay Mamata and Chakraborty Anuradha, *"Process for Sterilization of Biomaterials Using Supercritical Fluids"*, Indian Patent No. 211305, 2007, Application No. 543/MUM/2004.
- Mukhopadhyay Mamata and Chakraborty Anuradha, *"Lipoxygenase-Inactivated and Sterilised Legumes and Cereal Products"* Indian Patent No. 282732, 2017, Application No. 540 /MUM/2005.

- Region Global Forecast to 2025, www.marketsandmarkets.com, June 2019.
- Veryasova N. N, Lazhko A. E, Isaev D. E, Grebenik E. A, Timashev P. S, *"Supercritical Carbon Dioxide—A Powerful Tool for Green Biomaterial Chemistry"*, Russian Journal of Physical Chemistry B volume 13, pages 1079–1087, 2019.

❐

2

Classification of Foods, Biomaterials, and Microorganisms

2.1 Food Constituents

Food is any substance normally eaten or drunk by living beings. Food is the main source of energy and nutrition for humans, and is usually of animal or plant origin. A deep understanding and knowledge of the constituents of food and their properties are essential to food science. Any food system, i.e., fruits, vegetables, cereals, pulses, legumes, etc., is known for its specific characteristics, such as, water content, color, texture, aroma, taste, and nutritional value. These are discussed briefly in the following sub-sections. The constituents and the properties of the foods are important from the perspective of technology development. The specific constituents make each food system unique and the technology developed should be such that it does not negatively alter the original constituents and the inherent properties of the food being processed.

2.1.1 Water

In vegetal cells, water is present in the following forms:

i) Bound water or dilution water, which is present in the cell, forms true solutions with mineral or organic substances.

ii) Colloidal bound water which is present in the membrane, cytoplasm, and nucleus, acts as a swelling agent for these colloidal structured

M. Mukhopadhayay and A. Chatterjee, *Sterilization and Preservation*,
https://doi.org/10.1007/978-3-031-17370-7_2

substances. This colloidal bound water is very difficult to remove during drying/dehydration processes.

iii) Constitution water which is directly bound to the chemical component molecules, is also removed with difficulty.

Water is a major component of food; its presence can be anywhere between 10% (grains) to 50% (meat products) to 70-80% (fruit and vegetable products). However, it provides an ideal environment for bacterial growth and consequently food spoilage, if proper processing is not done. The amount of water present in a food system influences its textural properties, and determines the extent to which microbial spoilage may occur. One of the ways by which this is measured in food is by determining the water activity, which is very important for establishing the shelf life of processed foods. One of the basic principles of food preservation is to reduce the amount of water to enhance shelf life. Methods, such as, dehydration, freezing, refrigeration, etc., use this very principle. Removing water results in reduction of "free" water and prevents microbial growth. Generally, bacteria require a water activity of > 0.9 to grow and most yeasts and molds are inhibited by a water activity of < 0.7. The water activity (a_w) of a food is the ratio between the vapor pressure of the food itself, when in a completely undisturbed balance with the surrounding air media, and the vapor pressure of distilled water under identical conditions. A water activity of 0.80 means the vapor pressure is 80% of that of pure water. The water activity increases with temperature. The moisture condition of a product can be measured as the equilibrium relative humidity (ERH) expressed in percentage or as the water activity expressed as a decimal.

Most foods have a water activity above 0.95 and that will provide sufficient moisture to support the growth of bacteria, yeasts, and mold. The amount of available moisture can be reduced to a point which will inhibit the growth of the organisms. If the water activity of food is controlled to 0.85 or less in the finished product, it is not subject to the regulations of 21 CFR Parts 108, 113, and 114 *[US FDA, 2014]*.

2.1.2 Minerals

Mineral substances are present as salts of organic or inorganic acids or as complex organic combinations (e.g., chlorophyll, lecithin, etc.); they are in many cases dissolved in cellular juice. Vegetables are richer in mineral substances as compared to fruits. The mineral content is normally between 0.60 and 1.80% and more than 60 elements are present, the major elements being: K, Na, Ca, Mg, Fe, Mn, Al, P, Cl, and S. Important quantities of potassium (K) and absence of sodium chloride (NaCl) give a high dietetic value to fruits and to their processed products. Mainly vegetables supply phosphorus. Vegetables usually contain more calcium than fruits. Green beans, cabbage, onions, and beans contain more than 0.1% calcium.

Minerals may be broadly classified as macro (major) or micro (trace) elements. The macro minerals include calcium, phosphorus, sodium, and chloride, while the micro elements include iron, copper, cobalt, potassium, magnesium, iodine, zinc, manganese, molybdenum, fluoride, chromium, selenium, and sulphur. Macro minerals are required in amounts greater than 100 mg/day, while the micro minerals are required in lesser amounts.

2.1.3 Carbohydrates

Carbohydrates are the main constituents of fruits and vegetables, and represent more than 90% of their dry matter. Carbohydrates in foods are a combination of carbon, hydrogen, and oxygen and can be classified as simple and complex carbohydrates. The sugars constitute simple carbohydrates and complex carbohydrates are starches and fibers. From energy point of view, carbohydrates represent the most valuable food components. Daily adult intake should contain about 500 g carbohydrates. Carbohydrates can be oxidized to furnish energy, and glucose in the blood is a ready source of energy for the human body. Fermentation of carbohydrates by yeast and other microorganisms can yield carbon dioxide, alcohol, organic acids, and other compounds. Carbohydrates may be grouped under two kinds, namely, simple and complex:

i) Simple carbohydrates are classified as monosaccharides, disaccharides, trisaccharides, etc. Carbohydrates can be broadly divided into two types: (a) reducing and (b) non-reducing sugars. Examples of these are glucose (reducing) and sucrose (non-reducing). Reducing sugars contain a reactive aldehyde (CHO) or keto (C=O) group that is absent in non-reducing sugars. Thermal processing may trigger reactions between reducing sugars and the amino group of proteins, causing browning resulting in altering color and flavor of the products. This reaction is known as the Maillard's reaction. Similarly, heat processing, where high temperatures are used in a low water environment, can result in caramelization of sugars.

ii) Complex carbohydrates have a number of monosaccharide units linked by glycosidic linkages. Starch, cellulose, hemicellulose, pectin, etc., are examples of complex carbohydrates. Starch is made up of a number of glucose units linked by α-1-4 and α-1-6 glycosidic linkages that can be digested by humans. Starches that have 100% amylopectin are clear and do not form films. Starches that have >20% amylose, are pudding like, cloudy, and form films. Regular starches, when heated, replace the hydrogen bonds between starch molecules with starch-water bonds, causing gelatinization and imparting a thickening effect.

Sugars present in fruits and vegetables, such as, glucose, fructose, maltose, and sucrose all share the following characteristics in varying degrees:

i) They are readily fermented by microorganisms.

ii) They may be used as a preservative, as they prevent the growth of microorganisms in high concentrations.

iii) On heating, they darken in color or caramelize.

iv) Some of them combine with proteins to give dark colors, known as the browning reaction.

2.1.4 Fats and Lipids

Vegetables, and fruits, in general, provide low levels of fats, (less than 0.5%). However, considerable amounts are found in nuts (55%), apricot kernel (40%), grapes seeds (16%), apple seeds (20%), and tomato seeds (18%).

Lipids include a wide range of molecules, such as, water insoluble/non-polar compounds of biological origin. Triglycerides, fatty acids, phospholipids, sphingolipids, glycolipids, terpenoids, waxes, retinoids, and steroids, all fall under lipids. A triglyceride has three fatty acids which undergo esterification to form three hydroxyl groups of glycerol. Triglycerides containing unsaturated fatty acids are oils and those that contain mostly saturated fatty acids are fats. Owing to a higher level of unsaturated fatty acids, oils will oxidize over time. Hydrogenation is done to convert vegetable oils into semi-solid/solid fats for use in baked/processed foods. The partially hydrogenated products are less prone to oxidation than the oils. Lipids in food include oils of grains, such as, ground nuts, soybean, rice bran, sunflower, corn, and animal fats, such as, milk, cheese, ghee, and meat.

2.1.5 Organic Acids

Fruits contain natural acids, such as, citric acid in oranges and lemons, malic acid in apples, and tartaric acid in grapes. These acids give the fruits tartness and slow down bacterial spoilage. Organic acids influence the color of foods, since many plant pigments are natural pH indicators. With respect to bacterial spoilage, the most important contribution of organic acids is in lowering pH of the foods. Under anaerobic conditions and around a pH of 4.6, *Clostridium botulinum* can grow and produce lethal toxins. This hazard is absent in foods high in organic acids resulting in a pH of 4.6 and less.

2.1.6 Nitrogen Containing Substances

Nitrogen containing substances are found in plants as different combinations, such as, proteins, amino acids, amides, amines, nitrates, etc., e.g., vegetables contain 1.0 - 5.5 % of them. In fruits nitrogen-containing substances are less than 1% in most cases. Among nitrogen containing substances the most important ones are proteins. Proteins have a colloidal structure and heating their solution above 50°C makes them insoluble. This behavior has to be taken into account in heat processing of fruits and vegetables. [*Gibbs and Steele, 2020*].

2.1.7 Vitamins

Vitamins are defined as organic materials, which must be supplied to the human body in small amounts apart from the essential amino acids or fatty acids. Vitamins by themselves do not provide energy, although they may participate as coenzymes in chemical reactions which, in turn, release energy from other molecules. The vitamins are divided into two major groups: those that are fat-soluble and those that are water-soluble. The fat-soluble vitamins are A, D, E, and K. Their absorption by the body depends upon the normal absorption of fat from the diet. The water-soluble vitamins include vitamin C and several members of the vitamin B complex.

2.1.8 Proteins

Proteins are polymers of amino acids linked together through a peptide bond, mainly composed of carbon, nitrogen, hydrogen, oxygen, and sulfur. They may sometimes also contain iron, copper, phosphorus, and zinc. The function of a protein is determined by the sequence of its amino acids. There are twenty amino acids that are found in proteins, of which ten are called as essential amino acids, as these cannot be produced by the body and have to be provided through diet.

In foods, proteins add texture, contribute to the odor and taste, form gels, and stabilize foams and emulsions, etc. The conditions used in food processing are adjusted, such that it has optimal effect on proteins and other food characteristics. For example, in bread the brown crust is due to the Maillard's reaction while the final structure of the bread is caused by the thermal gelation of the protein gluten.

2.1.9 Enzymes

Food enzymes should be considered essential nutrients as they are capable of digesting food before the body's own digestive process begins and can enhance the transport of nutrients to the blood irrespective of a compromised digestive system. However, these are often removed from the raw foods during their processing to enhance shelf life.

For example, pineapples contain a group of digestive enzymes, called bromelain. These enzymes are *proteases,* which break down protein into its building blocks, including amino acids and aid in the digestion and absorption of proteins. Papaya is another tropical fruit that is rich in digestive enzymes. Like pineapples, papayas also contain *proteases* that help digest proteins. However, they contain a different group of *proteases,* known as, *papain*. Mangoes contain the digestive enzyme *amylase* which breaks down carbs from starch (a complex carb) into sugars like glucose and maltose. *Amylase* also helps mangoes ripen.

Raw honey contains a variety of digestive enzymes, including *diastase, amylase, invertase,* and *protease*. In processed honey these digestive enzymes are destroyed, if heated.

Bananas contain *amylases* and *glucosidases*, two enzymes that digest complex starches into easily absorbed sugars. They are more active as bananas start to ripen, which is why yellow bananas are much sweeter than green bananas.

Avocados contain the digestive enzyme *lipase* which breaks down fat molecules into smaller fatty acids and glycerol. That is why, consuming avocados may ease digestion after a high-fat meal. Kiwifruit contains the digestive enzyme *actinidain*, which helps digest proteins. Moreover, consuming kiwifruit may ease digestive symptoms like bloating and constipation. Ginger contains the digestive enzyme *zingibain* which is a *protease*. It may aid digestion by helping food move faster through the digestive tract and boosting the body's own production of digestive enzyme.

Enzymes are biological catalysts that promote most of the biochemical reactions that occur in cells. The properties of enzymes important in the technology for processing fruits and vegetables are:

(a) Enzymes control the reactions associated with ripening in living fruits and vegetables.

(b) After harvest, unless destroyed by heat, chemicals or some other means, enzymes continue the ripening process, in many cases to the point of spoilage, such as, soft melons or overripe bananas.

(c) Because enzymes enter into a vast number of biochemical reactions in fruits and vegetables, they may be responsible for changes in flavor, color, texture, and nutritional properties.

(d) The heating processes in manufacturing/processing of fruits and vegetables are designed not only to destroy microorganisms but also to deactivate enzymes and improve storage of the fruits and vegetables.

Enzymes need an optimal temperature around 50°C at which their activity is at maximum. Heating beyond this optimal temperature deactivates the enzyme. Activity of each enzyme is also characterized by an optimal pH. The enzyme classes like *hydrolases* (*lipase, invertase, tannase, chlorophylase, amylase, cellulase*) and *oxidoreductases* (*peroxidase, tyrosinase, catalase, ascorbinase, polyphenol oxidase*) play the most important roles in fruits and vegetable storage and processing. However, for food preservation and juice stabilization, some enzymes need to be deactivated to enhance shelf life.

Understanding the enzyme action and its structure is key to determining the optimal treatment conditions for inactivating undesirable enzymes. A slight tweak in the treatment conditions, such as, temperature, may act as an activator or a deactivator for enzymes; hence knowledge of the enzyme, its activity, and reaction rates, is imperative.

2.1.9.1 Pectin esterase

Pectin is a water-soluble gelatinous polysaccharide fiber which is present in ripe fruits and is used as emulsifier, stabilizer, gelling, and thickening agent in various foods, e.g., jams and jellies. Commercial pectins are extracted from citrus and apple fruit. *Pectin esterase (PE, EC 3.1.1.11)* removes methoxyl groupings from methylated pectin, as shown in *Figure 2.1*. This results in the formation of methanol and low methoxyl pectin. *PE* acts preferentially on methoxyl groups of esterified galacturonic acid molecules which are situated either at the non-reducing end of the polygalacturonan backbone chain or next to a molecule of non-esterified galacturonic acid.

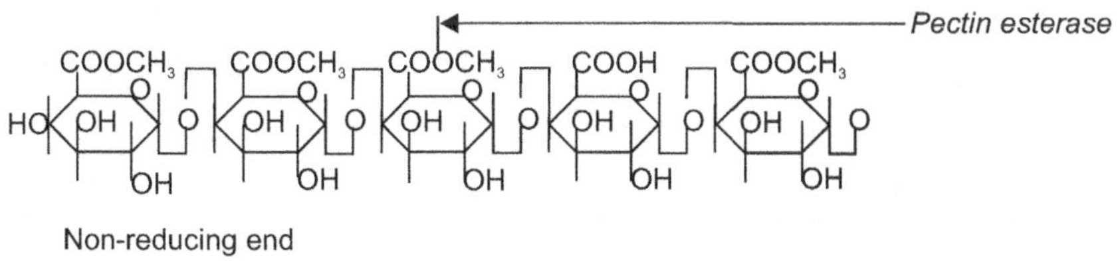

Fig. 2.1 : Degradation of pectin by the action of *Pectin esterase*. [*Voragen, 1991*]

The *PE* of plant origin proceeds along the polygalacturonan backbone in a sequence which creates regions of non-esterified galacturonic acid. This in itself will not reduce the viscosity of a solution of such pectin, as the degree of polymerization is not affected. In the presence of calcium ions the viscosity may even increase, as Ca^{2+} cross-links between long lengths of de-esterified galacturonic acid molecules in adjacent chains, as shown in *Figure 2.2*.

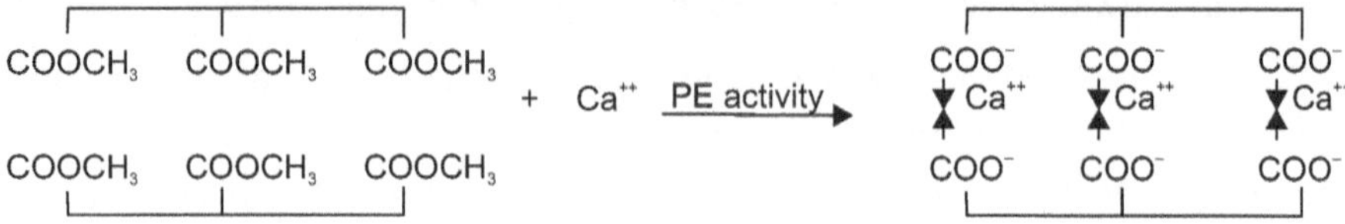

Fig. 2.2: Cross – linking of Ca^{2+} in presence of *PE* activity. [*Voragen, 1991*]

This activity results in binding the chains together, thus increasing the viscosity of the pectin solution. These enzymes have a pH optimum of 7.0 or above, and show stronger affinities for pectins with a lower degree of esterification.

Although, *PE* is useful in a number of ways, such as, fruit maturation, and photosynthetic metabolism of plant; however, from the perspective of fruit juice industry, it is undesirable, since demethylation of pectin, catalyzed by this *PE*, leads to separation of serum forming particulates in juice and imparting a cloudy appearance to the juice. Also, its presence adversely affects the delicate flavor of the juices inducing off-flavors *[Kohli et. al., 2015]*.

2.1.9.2 Polygalacturonase

Polygalacturonase (*PG, EC 3.2.1.15* endo and *EC 3.2.1.67* exo) exists in two forms: endo -*PG* and exo-*PG*. Both enzymes are *depolymerases* and act only on glycosidic linkages between galacturonic acid molecules, which are non-esterified. Endo-*PG* acts on the polygalacturonic backbone randomly, whereas, exo-*PG* only acts on the relevant bond at the non-reducing end of the chain, as shown in *Figure 2.3*.

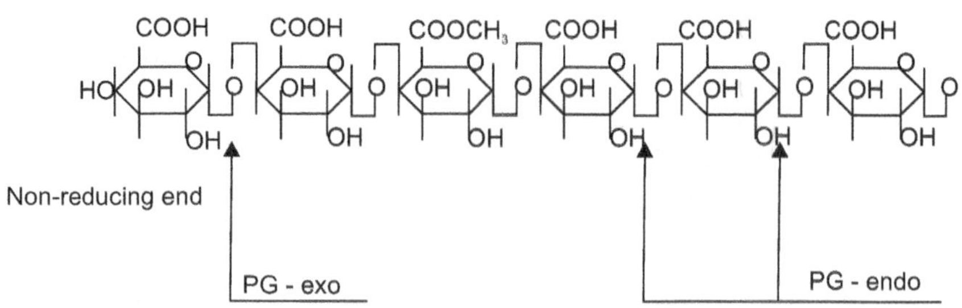

Fig. 2.3: Enzymatic action of *Polygalacturonase* on pectin. (*Voragen, 1991*)

Exo-*PG* would only sequentially release small fragments from the reducing end of the polysaccharide chains, and so would not greatly reduce the viscosity. Endo-*PG* would rapidly cause depolymerization reducing the viscosity. Endo-*PG* has a pH optimum of around 4.0-4.5, whereas exo-*PG* has a pH optimum of about 5.0 *[Voragen, 1991].*

The action of *PG* must be prevented to ensure that products are not cloudy with a high viscosity, and also to limit loss in tissue firmness *[Duvetter et al., 2009].*

2.1.9.3 Polyphenol oxidase

Polyphenol oxidase (PPO, EC 1.10.3.1), also known as *catechol oxidase*, and monophenol *monooxygenase (EC1.14.18.1)* are responsible for enzymic browning and often coupled to loss of vitamin C. The enzyme complexes from higher plants and fungi are virtually nonspecific and oxidize a wide variety of monophenolic and o-diphenolic compounds. Depending on its biological source, it may have monophenol *monooxygenase* and/or catechol *oxidase* activities. Monophenol *monooxygenase* (*tyrosinase*) can oxidize both tyrosine and dihydroxyphenylalanine (DOPA) to melanin, whereas *catechol oxidase* can convert only DOPA to melanin *[Nagodawithana and Reed, 1993]*. *PPO* is a Cu-containing enzyme which is associated with dark pigmentation; in plants, both soluble and membrane bound *PPO*'s have been described. Histochemical tests reveal that these are located in the chloroplasts *[Martinez* and *Whitaker, 1995].*

Polyphenol oxidase (*PPO*), the copper-containing metalloprotein which is involved in oxidation of phenol to quinone, is the major biochemical reaction that causes melanosis. This enzymatic action causes unappealing blackening in fruits, juices, and sea foods, making it unacceptable to consumers *[Mishra et al., 2016].*

2.1.9.4 Lipoxygenase

Lipoxygenases (*LOX, EC 1.13.11.12*) play an important role in the stability of various processed food products. All *lipoxygenases* use molecular oxygen (dioxygen) to catalyze the oxidation of polyunsaturated fatty acids and initially form fatty acid peroxy free radicals that remove hydrogen from another unsaturated fatty acid molecule to produce hydroperoxides, which can be broken down by other enzymes to form desirable and characteristic aroma compounds. The hydroperoxides are converted into acids, aldehydes or ketones, and other compounds. *Lipoxygenases* also can undergo an anaerobic reaction. Soybean *lipoxygenase*-1 can use the hydroperoxide product as its second substrate. Although the mechanism is simple, a complex pattern of product is formed: fatty acid dimers, oxodienoic acids, and n-pentane. *LOX* may also catalyze the co-oxidation of carotenoids, including β-carotene, resulting in the loss of essential nutrients and the development of off-flavors. *[Nagodawithana and Reed, 1993]. Figure 2.4* illustrates the mode of action for *LOX*.

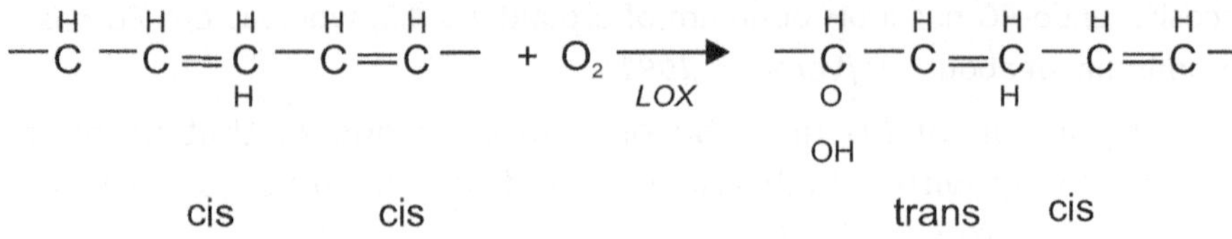

Fig. 2.4: Mode of action of *LOX [Prigge et al., 1997]*

2.1.9.5 Actions and Protein Structure of Enzymes

Enzymes work as catalysts in living cells and have the ability to accelerate/decelerate biochemical reactions without itself being altered in anyway during the process of the reaction. They are highly selective and are a part of many metabolic processes that occur in all organisms. Food will have presence of not just the naturally occurring enzymes, but also the enzymes produced by the microflora present in it. Enzymes can help to control ripening and changes in the flavor, color, texture, and nutritional value of different foods.

These enzymes are made up of one or more polypeptide backbones with a specific set of amino acid sequence and have specific folded arrangement. This natural folded structure of an enzyme is referred to as its native state. In effect, the primary structure of an enzyme is unique sequence of amino acids in a polypeptide chain. The specific folding of this polypeptide chain forming the β-sheet structure and α-helix is referred to as the secondary structure. These structures are held together by the hydrogen bonds of the amino acids. The 3-D arrangement of the polypeptides is the tertiary structure. Enzymes have an active site, where the reaction is catalyzed. This active site has a specific shape and specific functional groups that bind to the reacting molecules. The substrate molecule can bind to the active site via non-covalent interactions like hydrogen bonding, hydrophobic interactions, electrostatic interactions, and van der Waals forces. The structure formed by the union of more than one protein molecule is known as the quaternary structure. The functional arrangements, namely, the tertiary and quaternary structures, are called conformations.

Figure 2.5 presents enzyme (protein) structure and different levels of protein organization. A change in this native conformation is known as conformational change. A conformational change may result in partial or complete inactivation of the enzyme, no change in activity, or trigger enzyme activation. The enzyme conformation is directly affected by pH, temperature, pressure, salt concentration, and chemical or physical stresses, which can cause denaturation of the protein. Denaturation can be reversible if the primary structure of the polypeptide is conserved in the process, as the protein regenerates and resumes its function. Denaturation is irreversible, leading to partial or total enzyme inactivation, when its native conformation cannot be conserved. The deteriorative enzyme activity is not desirable in food, as it affects the food characteristics, such as, flavor, taste, shelf life, and aroma. The action of enzymes has beneficial effects as well. Examples of enzymes used in the food industry include *proteases*, which are used to help tenderize meat. *Lipase* is known to help provide flavor to foods, such as, chocolate and cheese, and *amylase* converts starches to sugars in industries, such as, brewing and baking *[Silva et al., 2017]*.

2.2 Classification of Foods by Characteristic Properties

2.2.1 Texture

The range of textures that are encountered in fresh and cooked vegetables and fruits is wide and to a large extent can be described in terms of changes in specific cellular components. Since more than two-thirds of plant tissues are generally water, the relationships between these components and water further determine textural differences. The state of turgidity, as determined by osmotic forces, plays a paramount role in the texture of fruits and vegetables. The cell walls of plant tissues have varying degrees of elasticity and are permeable to water, ions, and small molecules. The membranes of the living protoplast are semi-permeable, that is, they allow passage of water but are selective with respect to transfer of dissolved and suspended materials. The cell vacuoles contain most of the water in plant cells and sugars, acids, salts, amino acids, some water-soluble pigments, and vitamins, and other low molecular weight constituents are dissolved in this water. In the living plant, water taken up by the roots passes through the cell walls and membranes into the cytoplasm of the protoplasts and into the vacuoles to establish a state of osmotic equilibrium within the cells. The osmotic pressure within the cell vacuoles and within the protoplasts pushes the protoplasts against the cell walls and causes them to stretch slightly in accordance with their elastic properties. This is (in growing plant and the harvested live fruit or vegetable) responsible for desired plumpness, succulence, and much of the crispness. When plant tissues are damaged or destroyed by storage, freezing, cooking, or other causes, an important major change that results is denaturation of the proteins of cell membranes resulting in the loss of perm-selectivity. Without perm-selectivity the state of osmotic pressure in cell vacuoles and protoplasts cannot exist, and water and dissolved substances are free to diffuse out of the cells and leave the remaining tissue in a soft and wilted condition.

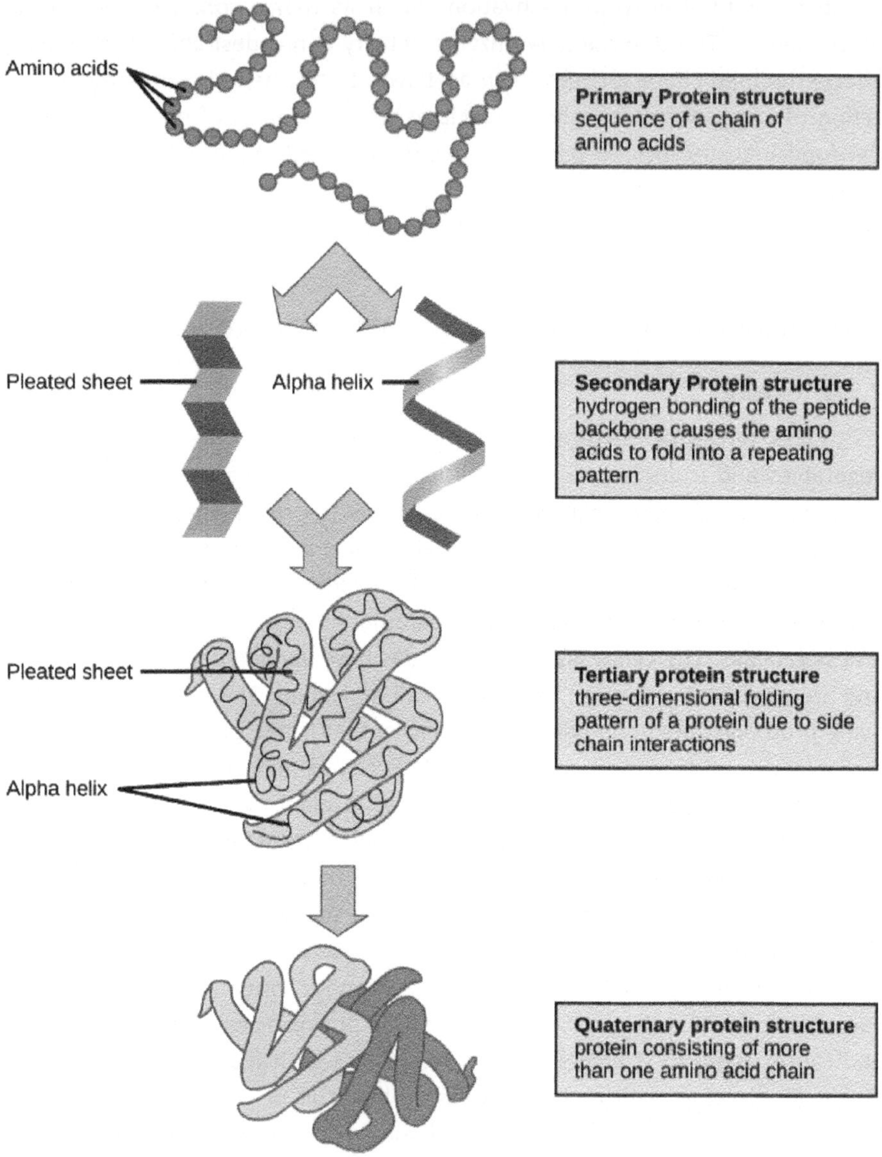

Fig 2.5: Enzyme (Protein) structure. *[National Institutes of Health, 2019]*

Cell walls in young plants are very thin and are composed largely of cellulose. With the ageing of the plant, the cell walls thicken and become higher in hemicellulose and lignin. These materials are fibrous and tough, and are not significantly softened by cooking. The complex polymers of sugar acid derivatives include pectin and closely related substances. The cement-like substance found especially in the middle lamella, which helps plant cells to hold to one another, is a water-insoluble pectic substance. On mild hydrolysis it yields water-soluble pectin that can form gels or viscous colloidal suspensions with sugar and acid. Certain water-soluble pectic substances also react with metal ions, particularly calcium, to form water-insoluble salts, such as, calcium pectates. The various pectic substances may influence texture of vegetables and fruits in several ways. When vegetables or fruits are cooked, some of the water-insoluble pectic substances are hydrolyzed into water-soluble pectin. This results in a degree of cell separation in the tissues and contributes to tenderness. Since many fruits and vegetables are somewhat acidic and contain sugars, the soluble pectin also tends to form colloidal suspensions, which will thicken the juice or pulp of these products. Fruits and vegetables also contain a natural enzyme, which can further hydrolyze pectin to the point where the pectin loses much of its gel forming property. This enzyme is known as *pectin methyl esterase*. Materials, such as, tomato juice or tomato paste will contain both pectin and *pectin methyl esterase*.

If freshly prepared tomato juice or paste is allowed to stand, the original viscosity gradually decreases due to the action of *pectin methyl esterase*. This can be prevented if the tomato products are quickly heated to a temperature of about 82°C, in order to deactivate the *pectin methyl esterase* liberated from broken cells, before it has a chance to hydrolyze the pectin. Such a treatment is commonly practised in the manufacture of tomato juice products. This is known as the "hot-break process" and yields products of high viscosity. In contrast, where low viscosity products are desired, no heat is used and enzyme activity is allowed to proceed. This is "cold-break" process. After sufficient decrease in viscosity is achieved, the product can be heat treated, as in canning, to preserve it for long-term storage. It is often also desirable to firm the texture of fruits and vegetables, especially when products are normally softened by processing. In this case, advantage is taken of the reaction between soluble pectic substances and calcium ions that form calcium pectates. These calcium pectates are water insoluble and when they are produced within the tissues of fruits and vegetables, they increase structural rigidity. Thus, it is common commercial practice to add low levels of calcium salts to tomatoes, apples, and other vegetables and fruits prior to canning or freezing.

2.2.2 Color

Color provides unique characteristics to fruits and vegetables. Consumers' acceptance of a particular processed product also depends on the color and consistency. The pigments and color precursors of fruits and vegetables occur for the most part in the cellular plastid inclusions, such as, the chloroplasts and other chromoplasts, and to a lesser extent dissolved in fat droplets or water within the cell protoplast and vacuoles. These pigments are classified into four major groups, which include the chlorophylls, carotenoids, anthocyanins, and anthoanthins. Pigments belonging to the latter two groups are also referred to as flavonoids and include the tannins.

2.2.3 Flavor

As defined by Anon in 1959, flavor is a mingled but unitary experience which includes sensations of taste, smell, and pressure, and often cutaneous sensations, such as, warmth, color, or mild pain. Flavor typically encompasses aroma/odor and taste. Aroma compounds are volatile while taste receptors exist in the mouth which respond when the food is chewed. In the evaluation of fruit and vegetable flavor, the consideration of off-flavors as well as desirable ones is key. The off-flavors may be a result of the enzymatic action, such as, *lipoxygenase* or *peroxidase*; these produce reactive free radicals and hydroperoxides that may catalyze the oxidation of lipid compounds. These reactions may result in development of undesirable flavors described as rancid, cardboard, or oxidized. However, there may be exceptions wherein the reactions cause desirable flavors.

2.2.4 Nutritional Aspects

Nutritional value refers to contents of food and the impact of constituents on body. The nutritive aspects of food need to be considered from a dual point of view: first, a study of nutrients that a particular food contains and subsequent evaluation of need of these nutrients in human. Second, a study is focused on the relative stability of these nutrients and the effects of food processing, storage, and preparation on these nutrients. The nutrients in food, required in balanced amounts, belong to the broad groups of carbohydrates, proteins, fats, vitamins, and minerals. These have already been discussed in the previous paragraphs. The science of nutrition, also deals with the physiological and biochemical phenomena of food utilization as related to health.

2.2.5 Food Extracts and Phyto-Pharmaceuticals

Phyto-pharmaceuticals may be defined as plant-based enriched fractions of anthocyanidins, carotenoids, lycopenes, flavonoids, glucosinolates, isoflavonoids, limonoids, polyphenols, omega-3 fatty acids, phytoestrogens, resveratrol, phytosterols, probiotics, and terpenoids, which have specific pharmacological effects on human health. They usually possess therapeutic properties against inflammation, allergies, oxidation, and microbial infections, diabetes, ageing, etc. A drug made from phyto-pharmaceuticals is a purified and standardized fraction with a minimum of four bioactive or phytochemical compounds extracted from a medicinal plant or its part, for internal or external use, but it cannot be administered via parenteral route.

The methods of extracting phyto-pharmaceuticals permitted by the Pharmaceutical Regulatory body for drug development technologies include solvent extraction, fractionation, potentiating steps, add-back techniques, modern extraction techniques (like CO_2-based extraction), freeze-drying, formulation developments, etc., *[Nooreen et al., 2018]*.

As per a market analysis, the plant extracts market is estimated to be valued at USD 23.7 billion in 2019 and is projected to reach USD 59.4 billion by 2025. A rise in awareness regarding the health benefits offered by phyto-pharmaceutical extracts, has significantly fuelled the market for plant extracts adding to the growing need for plant extracts in food and beverage industry *[Markets and Markets, 2019]*.

2.2.6 Safety, Diversity and Complexity

Food safety is defined as the inverse of food risk by the probability of not suffering some hazard from consuming a specific food *[Henson and Traill, 1993]*. Potential sources of undesirable contaminants in foods may be widely varied from natural (e.g., mycotoxins), environmental (e.g., dioxins), agro-chemicals (e.g., nitrates and pesticides), growth promoters, to packaging components, to name a few. Microbiological contaminants pose even greater challenges to safety of food owing to potentially harmful microorganisms, which have the ability either to grow rapidly from very low numbers in food during storage or to proliferate in the human body once ingested. Food systems are integrally related to food safety. Contamination can occur at any point in the food system. Since food is directly related to the well-being of people, it is pertinent to have stringent prevention and control strategies that can be implemented at any point.

The limit of food safety is often indicated by the term shelf life which is defined as the finite length of time a processed food product can be stored under specific packaging or environmental condition whilst still maintaining an acceptable quality or specific functionality. A product does not immediately become dangerous for human consumption, even after the shelf life has crossed and so shelf life does not always correspond to food safety. Some food products may remain fresh for days even after the shelf life has expired provided there is no bacterial contamination, e.g., pasteurized milk. So proper packaging of processed foods is important for food safety.

Food systems have been constantly evolving, with each change bringing new advantages and challenges. The evolution of food systems has created ever-growing diversity and complexity arising out of wide variations in the availability of processed foods, such as, packaged food, precooked food, 'ready-to eat' food, fruit juices, fruit and spice pastes, and food extracts. The scale and complexities of today's food systems contribute to the likelihood and magnitude of food-borne illnesses. The more the complexity is, the more avenues are opened up for things to go wrong; the larger the reach, the more people will be potentially affected.

2.3 Factors Causing Degradation and Spoilage of Foods

Food spoilage is a process or change which occurs over a period of time, rendering the food product undesirable or unacceptable for consumption. It is a complex ecological phenomenon which occurs due to the biochemical activity of microbes which eventually dominates in response to the prevailing environment or ecological determinants. Food spoilage could be due to microbiological, chemical, or physical factors. Microbiological food spoilage is a result of microbial growth wherein microbes produce enzymes that lead to objectionable by-products in the food. Chemical food spoilage is a result of reaction of different components of food with each other or with the component that has been added to it, which alters the food's sensory characteristics. Physical food spoilage occurs when moist foods are dehydrated excessively or dry foods absorb excessive moisture or for overheating and charring or unwanted caramelization.

2.3.1 Physical Factors

Physical spoilage could be due to multiple factors, such as, physical damage to food during harvesting, processing or distribution, attack by rodents, insects and pests, unfavourable climatic conditions; for example, raining during off-season will spoil the growth of mangoes. The physical damage increases the probability of a subsequent chemical or microbial spoilage. Physical spoilage also increases the chances of contamination, since the protective outer layer of the food is ruptured and microorganisms gain an easy entry into the food.

2.3.2 Chemical Factors

The changes in the color and flavor of foods during processing and storage are due to the chemical reactions taking place within the food. Over a period of time, fats break down setting off rancidity, naturally occurring enzymes promote chemical changes in foods as they age, causing off-flavors. Enzymes play an important role in the decomposition of once-living tissues. The process is known as autolysis or enzymic spoilage, thus reducing the time between post-harvest/animal slaughter to processing and is the key to increasing the shelf life of the product.

Also, once the cells of fruits and vegetables, such as, apples, potatoes, bananas, and avocado are cut and exposed to air, enzymes present in them promote chemical reaction in which colorless compounds are converted into brown-colored compounds (enzymic browning). Processing of the food immediately after cutting inactivates the enzymes halting the browning effect. Non-enzymatic browning or Maillard's reaction (MR), which occurs between reducing carbohydrates and amines, is of crucial importance in food science, since it impacts the taste, aroma, and color of food substance.

2.3.3 Microorganisms

Pathogens are food-borne microorganisms, mainly bacteria, fungi, fungal spores, viruses, or even parasites that are present in the food and are the cause of major diseases, such as, food poisoning. Bacterial contamination is a common cause of food poisoning and food spoilage. Food poisoning occurs when disease-causing bacteria infest food and are consumed. In conditions ideal for bacterial growth, one single-cell bacteria can become two million in just seven hours.

Enterohemorrhagic *Escherichia coli* (*EHEC*) serotype *O157:H7* is a human pathogen responsible for outbreaks of bloody diarrhea and hemolytic uremic syndrome (HUS) worldwide. *EHEC* infection was the largest recorded outbreak of a bacterial infection observed in Germany in 2011. The huge number of HUS cases made it the largest outbreak of HUS worldwide. This food borne outbreak thus revealed how rapidly a food borne pathogen can spread and cause serious illness and death *[Hueston and McLeod, 2012]*.

Certain types of bacteria also produce bacterial toxins in the process of multiplying, thus producing waste. Bacterial toxins are dangerous botulinum; the bacterial toxin that causes botulism is the most potent natural poison known. The bacterium *Listeria monocytogenes* is found in soft cheeses, unpasteurized milk, meat, and seafood, and causes a disease called listeriosis. Food infection caused by *Escherichia coli* is colitis, *Campylobacter jejuni* causes the disease campylobacteriosis, shigellosis is caused by *Shigella* and *Staphylococcus aureus* causes staphylococcal food poisoning.

The bacteria of interest to food microbiology can be analyzed in terms of infectious agents, causes of foodborne intoxication, spoilage, and processing aids.

Table 2.1 Examples of bacteria of concern to food microbiology *[Gill, 2017]*

Food borne infectious agents	Foodborne intoxicants	Spoilage agents	Aids Processing
Brucella	*Bacillus cereus*	*Acinetabacter*	Lactic Acid Bacteria
Campylabacter	*Clostridium batulinum*	*Aicaligens*	*(Lactobacillus, Lactacoccus,Pediacoccus, Leucanastac, Streptococcus)*
Clostridium perfringens	*Clostridium perfringens*	*Bacillus*	Yeast
Cranobacter	*Staphylacaccus aureus*	*Brachothrix thermasphacta*	
Shigella		*Clostridium*	
Salmonella enterica		*Carnebacterium*	
Yersinia enteraclitica		*Enterabacteriaceae*	
Listeria monocytogens		*Erwinia Caratavora*	
Mycabacterium		*Moraxellaceae*	
Vibria		*Pseudomanas*	
		Shewanella Putrefacens	
		Vibria	
		Pediococcus spp.	
		Kurthia zopfiienus	

2.4 Undesirable Enzymes and Hurdles in Food Industry

Enzymes also have deteriorative effects causing food spoilage. Enzyme reactions causing the browning of foods, such as, apple, pear, peach, potato, lettuce, mushroom, and bananas are examples of deteriorative action of enzymes. Enzymes that are known to cause food spoilage include *lipase* which causes discoloration of cereals, promotes hydrolytic rancidity in oils, and imparts off-flavor to milk. Also, ascorbic acid oxidizes which is responsible for the loss of vitamin C in vegetables. Other deteriorative enzymes are *pectinesterase* and *poly phenol oxidase* which promote the softening and browning of fruits, and *protease* which causes reduction of gluten formation in flour. The major factors useful in controlling enzyme activity are: temperature, water activity, pH, and chemicals which can inhibit enzyme action, alteration of substrates, alteration of products, and pre-processing control.

Inactivation of undesirable enzymes is of importance to ensure the product quality and shelf life. The commonly used processing technologies for enzyme inactivation result in loss of nutritional value and quality deterioration. Thus, there is a need for alternative technologies to reduce the enzyme activity without modifying the food product.

Lipase has the ability to hydrolyze fats by splitting the fatty acid molecules from the glycerol molecule. The extent to which hydrolysis has occurred can be used as a measure to determine the *lipase* activity, by measuring the free fatty acid content. Ideally, a good quality product, should not exceed 1.2% of free fatty acid content which is expressed as oleic acid in the extracted fat.

2.5 Classification of Biomaterials by their Activity

Biological materials are materials that are produced by living organisms, such as, blood, bone, proteins, muscle, and other organic material. Biomaterials, on the other hand, are materials which are created specifically to be used for biological applications. Biomaterials may be natural or artificial (synthetic) and are used in medical applications to support, enhance, or replace damaged tissue or a biological function. The modern field of biomaterials combines medicine, biology, physics, and chemistry, and more recent influences from tissue engineering and materials science.

Biomaterials can be broadly classified in two categories, as shown in *Figure 2.6.*

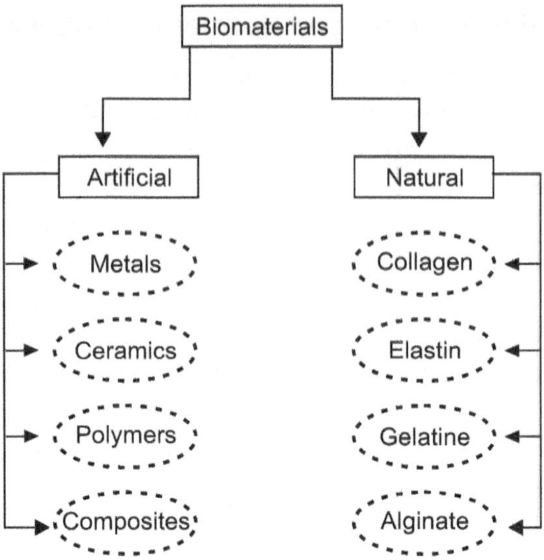

Fig. 2.6: Classification of Biomaterials based on materials

2.5.1 Natural Polymers

Natural biomaterials are any material taken from plants or animals and used to augment, replace, or repair body tissues and organs, e.g., bioceramics, coral, shells, biopolymers, etc. The use of natural biomaterials is not a new concept; naturally derived materials have been used by humans for thousands of years.

Different types of bioactive compounds are produced by plants. These include lectins, cyclotides, defensins, ribonucleolytic proteins, thionins, etc. Biomaterials extracted from plants have found varied applications in medical industry, e.g., for wound management and drug delivery as medical fibers and patches respectively. Natural products derived from insects are yet another potential area for the development of future biomaterials. Also, silk produced by silkworm, *Bombyx mori*, has excellent properties, such as, biocompatibility, biodegradation, nontoxicity, and adsorption properties. It has been commercially used as a biomaterial suture for decades and its application also includes wound management, enzyme immobilization matrices, vascular prostheses, and structural implants among others. Research into a variety of antimicrobial peptides, such as, megamins, defensins, cathelicidins, and protegrins generated by vertebrates, has become popular over recent times.

Biopolymers are obtained from living tissues or living organisms owing to their polymeric structural properties that exhibit properties, such as, strength, steadiness, and flexibility. Living organisms could include parts of plants or animals or microorganisms, which are used to produce biopolymers. These biopolymers have the ability to immediately overlap and acquire different shapes with complex delicate structures. Biopolymers of plant origin, such as, cellulose, pectin, hemicellulose, and lignin have gained immense industrial importance. The marine fauna and flora also contribute significantly towards providing biopolymeric ingredients. Biopolymers are usually classified on the basis of their biological nature, such as, proteins, carbohydrates, lipid wax, polyphenols, nucleic acids, and polyhydroxyalkanoates (PHA). *Table 2.2* and *Table 2.3* provide examples and uses of biopolymers of plant as well as animal origin *[Aggarwal et al., 2020]*.

Biopolymers too have found extensive use in the medical industry for augmentation or build-up of dermal tissue for cosmetic reasons, drug delivery systems, and haemostatic device, sutures. In fact, chitosan is used as an intraocular lens material because of its oxygen permeability and it has also been found to expedite blood clotting *[Ige et al., 2012]*.

Table 2.2: Biopolymers of Animal Origin [*Ige et al., 2012*]

Biopolymer	Origin	Application
Hyaluronic acid	The umbilical cord of new born child, rooster combs, fermentation broths of streptococcus and other bacteria	Gel preparation for drug delivery, wound healing, cosmetic products, filler in medicine, antibacterial.
Chitosan	Shellfish and crustacean waste materials	Cosmetic industry, medicine, wound healing treatment, food packaging, personal hygiene products, anti-bacterial, drug carrier for controlled release, bioremediation of toxic phenolic compounds, promote osteogenesis, fat absorbent action.
Gelatin	Cattle hides, bones, fish, pig skins	Pharmaceutical and medical usage.
Keratin	Feathers, hair, nails, wool, horn and hooves, stratum corneum	Drug delivery system, surgery, food industry, cosmetics, biomedical products.
Collagen	Invertebrates in the body walls and cuticles	Sutures, dental composites, skin, regeneration templates, cosmetics, biodegradable matrices.

Table 2.3: Biopolymers of Plant Origin [*Ige et al., 2012*]

Biopolymer	Origin	Application
Cellulose	Plant tissues, Bacteria	Controlled Drug Delivery devices, Wound Dressings, Scaffolds for Regenerative Medicine
Pectin	Plant cell walls, citrus peels, apple pomace	Reduce blood cholestrol, treat gastro-intestinal disorders, remove metals, such as, lead, mercury from intestine and lungs, control haemorrhage, tablet formulations, antimicrobial action, improves coagulation, treatment of over-eating, anti-inflammation
Carrageenan	Cell wall matrix of red seaweeds	Anticoagulant and antithrombotic activity, Antiviral activity
Xylan	Hardwood (Eucalyptus globules etc.) almond shell, rice husk, corn cobs	Low-caloric sweetener, preventative agent, drug delivery system
Gum Arabic	Stems and branches of *Acacia seyal* and *Acacia senegal* tree	Beauty products, medicine manufacturing industries, treatment of internal as well as external inflammation, antioxidant, nephroprotectant

2.5.2 Artificial Polymers

Artificial biomaterials are classified as: metals, ceramics, nonbiodegradable polymers, and biodegradable polymers. Some synthetic biomaterials are commercialized and applied in clinical treatment, such as, metal hip, acrylic or silicone intraocular lens. Classification may also be based on morphology, structure, composition, material strength of biomaterials (e.g., soft or hard tissues), and medical devices, made out of synthetic and natural polymers *[Soares et al., 2019; Vaisanen et al., 2017; Tomasko et al., 2003]*, and sensitive biomaterials (e.g., graft tissues) *[Lim et al., 2015; Ebara et al., 2014]* and nanomaterials *[Jon, 2018]*.

When a synthetic material is placed within the human body, tissue reacts towards the implant in a variety of ways depending on the material type. The mechanism of tissue interaction (if any) depends on the tissue response to the implant surface. In general, there are three terms in which a biomaterial may be described or classified according to the tissues responses. These are bioinert, bioresorbable, and bioactive.

The term bioinert refers to any material that once placed in the human body has minimal interaction with its surrounding tissue. Examples of these are stainless steel, titanium, alumina, partially stabilized zirconia, and ultra-high molecular weight polyethylene. Generally, a fibrous capsule might form around bioinert implants. Hence its bio-functionality relies on tissue integration through the implant.

Bioactive refers to a material, which upon being placed within the human body, interacts with the surrounding bone and in some cases, even soft tissue. This occurs through a time–dependent kinetic modification of the surface, triggered by their implantation within the living bone. An ion–exchange reaction between the bioactive implant and surrounding body fluids, results in the formation of a biologically active carbonate hydroxy apatite (CHAp) layer on the implant that is chemically and crystallographically equivalent to the mineral phase in bone. Prime examples of these materials are synthetic hydroxyapatite [$Ca_{10}(PO_4)_6(OH)_2$], glass ceramic A-W, and bioglass.

Bioresorbable refers to a material that upon placement within the human body starts to dissolve (resorbed) and slowly replaced by advancing tissue (such as, bone). Common examples of bioresorbable materials are tricalcium phosphate [$Ca_3(PO_4)_2$] and polylactic–polyglycolic acid copolymers. Calcium oxide, calcium carbonate, and gypsum are other common materials that have been utilized during last three decades.

2.5.3 Medical Implants

The field of medical devices has grown significantly in the past decade due to discoveries in tissue engineering, regenerative medicine, and more. Metals, ceramics, plastic, glass, and even living cells and tissue all can be used in creating a biomaterial. They can be reengineered into molded or machined parts, coatings, fibers, films, foams, and fabrics for use in biomedical products and devices. These may include heart valves, hip joint replacements, dental implants, or contact lenses. They often are biodegradable, and some are bio-absorbable, meaning they are eliminated gradually from the body after fulfilling a function. Some examples of medical devices are:

- Medical implants, including heart valves, stents, and grafts; artificial joints, ligaments, and tendons; hearing loss implants; dental implants; and devices that stimulate nerves.
- Sutures, clips, and staples for wound closure, and dissolvable dressings.

- Regenerated human tissues, using a combination of biomaterial supports or scaffolds, cells, and bioactive molecules. Examples include a bone regenerating hydrogel and a lab-grown human bladder.
- Molecular probes and nanoparticles that break through biological barriers and aid in cancer imaging and therapy at the molecular level.
- Biosensors to detect the presence and amount of specific substances and to transmit that data. Examples are blood glucose monitoring devices and brain activity sensors.
- Drug delivery systems that carry and/or apply drugs to a disease target. Examples include drug-coated vascular stents and implantable chemotherapy wafers for cancer patients.

It has been accepted that no foreign material placed within a living body is completely compatible. The only substances that conform completely are those manufactured by the body itself (autogenous) and any other substance that is recognized as foreign, initiates some type of reaction (host-tissue response).

Biomaterials have become more versatile with technological advancements such that it finds utility across the healthcare sector. The global biomaterials market size was estimated at USD 106.5 billion in 2019 and is expected to reach USD 121 billion by 2021. The rise in the number of cases of musculoskeletal and chronic skeletal medical conditions is expected to boost the demand for biomaterial-based implants, thereby augmenting the market growth. Also, there is a steady rise in the geriatric population who are at an increased risk of osteoarthritis, osteoporosis, and other musculoskeletal disorders, which, in turn, has led to an increase in the demand for orthopaedic implants *[Grand View Research, 2020]*.

2.6 Microorganisms Involved in Contamination of Biomaterials

Artificial materials, biomaterials, and biomedical goods lack the natural defence mechanism against infectious agents, as compared to living organisms. Microbial cells take advantage of this inherent lack of protection mechanism and adhere to any type of artificial surface, particularly in a moist environment and reproduce to form a huge population. For example, catheters being used for a long-time can lead to implant-associated infections. Nearly half of all nosocomial infections are related to the use of medical implants; these infections can be serious and even fatal. A contaminated surface immediately becomes a thriving ground for biofilm formation. These biofilms formed by microbial cells are basically polysaccharide matrix with embedded cells. Microbes form these biofilms as a protective mechanism aiding their survival under optimal conditions. These biofilms act as a barrier to antibiotics and biocides, making the cells less susceptible compared to microbial cells without the formation of biofilms. Pathogenic and resilient infections spread via the biofilms, wherein several toxins secreted by the microbial cells are accumulated in high concentrations in a closed system. The contamination due to medical devices depend on many factors, such as, the chemistry of the biomaterial, the physical properties of the surface, the design of the medical device, the anatomical site, the extension of surgical invasion, and the time of application. The host defence mechanism also contributes to the development of colonization and infection *[Alarfaj et al., 2016]*.

The clinical implication of using biomaterials is that there is a high chance of microbial colonization, the consequences of which can be varied. For example, intra-vascular catheter-related infections have been found to be a major cause of morbidity and mortality in the United States. Medical literature reports that >2% of about 200,000 primary hip prostheses and 200,000 primary knee arthroplasties, which are performed each year in the US, become infected within ten years *[Alarfaj et al., 2016]*.

2.6.1 Sources of Contamination

Viruses, fungi, protozoans, and bacteria are known to be involved in contamination of medical device and biomaterials. Water pipelines in surgical and dental units are suspected to be a potential transmission pathway of the hepatitis virus, as well as *Pseudomonas aeruginosa* and *Legionella*. Fungi such as, *Candida albicans* and *Aspergillus* are related to the early onset of prosthetic valve endocarditis right after implantation. Protozoans *Acanthamoeba*, known to colonize contact lens, are responsible for keratitis, a rare but devastating cause leading to loss of vision. *Table 2.4* enlists the most frequently found microorganisms in human related to medical device infection.

Table 2.4: Micro-organisms that are most frequently recovered in human infection and in infection-related devices *[Rimondini et al., 2005]*

Dental Caries	Acidogenic Gram-positive cocci (Streptococcus spp. and Lactobacilli
Muscoskeletal infections	*Staphylococci spp.*
End of Pregnancy/Child birth	*Clostridium*
Urinary tract/Woound infections	*Enterococcus Faecalis*
Native valve endocarditis	*Steptococcus viridans*
Cystic fibrosis pneumonia	*Pseudomonas aeruginosa, Burkholderia cepacia*
Meloidosis	*Pseudomonas pseudomallei*
Nosocomial Infection	
ICU pneumonia	Gram-negative rods
ICU	*Acinetobacter baumannii*
IV Solutions	*Burkholderia cepacia*
Catheter	*Carbapenem-resistant Enterobacteriaceae*
Breathing tubes/Catheter/ IV Solutions	*Stenotrophomonas maltophilia*
Sutures	*Staphylococcus epidermidis, Staphylococcus aureus*
Exit sites	*Staphylococcus epidermidis, Staphylococcus aureus*
Arteriovenous shunts	*Staphylococcus epidermidis, Staphylococcus aureus*
Scheletal buckels	Gram-positive cocci
Contact lens	*P. aeruginosa,* Gram-positive cocci

2.6.2 Classification of Agents of Contamination

The factors for contamination of biomaterials may be classified based on prions, endotoxins, and the biological strains of microorganisms, such as, bacteria, fungi, fungal spores, and virus.

2.6.2 1 Prions

The term prions refer to abnormal pathogenic agents that are transmissible and are able to induce abnormal folding of specific normal cellular proteins called prion proteins that are found most abundantly in the brain. This protein consists of about 250 amino acids. Some researchers believe that the prions are formed when PrP (Platelet-rich plasma) associates with a foreign pathogenic nucleic acid. This is called the virino hypothesis. Viruses consist of proteins and nucleic acids that are specified by the virus genome.

Prions are virus-like organisms made up of a prion protein. These elongated fibrils (green) are believed to be aggregations of the protein that makes up the infectious prion. Prions attack nerve cells producing neurodegenerative brain disease. Once prions infect the body, they cannot be destroyed. As they accumulate, the misshapen proteins somehow trigger neighbor proteins to behave similarly, eventually taking the place of normal proteins and destroying brain cells. Researchers have discovered the bacterial enzyme *keratinase* on brain tissues from cows with Bovine Spongiform Encephalopathy (BSE) or "mad cow disease" and sheep with scrapie that can degrade the prion protein and are held responsible for the disease *[Langeveld et al., 2003]*. When the tissue was pretreated and in the presence of a detergent, the enzyme was fully degraded in the prion, rendering it undetectable.

2.6.2.2 Endotoxins

Endotoxins are part of the outer membrane of the cell wall of gram-negative bacteria. Endotoxin is invariably associated with gram-negative bacteria irrespective of whether the organisms are pathogenic or not. Lipopolysaccharides (LPS), also known as endotoxins, are large molecules consisting of a lipid and a polysaccharide composed of O-antigen, outer core and inner core joined by a covalent bond; they are found in the outer membrane of gram-negative bacteria. Although the term endotoxin is occasionally used to refer to any cell-associated bacterial toxin, in bacteriology it is properly reserved to refer to the lipopolysaccharide complex associated with the outer membrane of gram-negative pathogens, such as, *Escherichia coli, Salmonella, Shigella, Pseudomonas,* and *Neisseria*. Endotoxins are dangerous; on entering the blood, they cause fever and a wide range of other possible effects including aseptic shock and death. The bacterial endotoxins come from the external membrane of gram-negative bacteria. These bacteria are characterized by their ubiquity in nature. They can be found in marine environments and on land, as well as, in the animals that some humans use for food, and in the feces of animals.

2.7 Contamination from Clinical Waste and Microflora

In general, the medical wastes may be classified under four different categories: infectious, hazardous, radioactive, and general. The infectious waste is any waste that poses the threat of infection to humans. This can include human/animal tissue, blood-soaked bandages, surgical gloves, cultures, stocks, or swabs that were used to inoculate cultures. Some infectious waste can even be labeled as pathological, which is any waste that could contain pathogens.

In response to the worldwide concern (and panic, especially during the unprecedented times of COVID) over medical waste, bio-medical waste management rules were formulated. It has been observed and strongly recommended that medical waste should be removed from the hospital within 24 hours of its generation to prevent environmental contamination caused by any accidental spillage of waste *[Saini et al., 2004]*.

A rational approach to managing medical waste is to identify wastes that represent a potential risk of causing infection while handling and disposing, in order that, for such wastes sufficient precautions need to be taken. Medical wastes include microbiology laboratory waste (e.g., microbiologic cultures and stocks of microorganisms), pathology and anatomy waste, blood specimens from clinics and laboratories, blood products, and other body-fluid specimens. All of these are highly infectious and they have the potential to cause an epidemic if stringent measures are not taken.

Regulatory bodies of each country have specific categories of medical waste and defined regulations and outlines for treatment and subsequent disposal. The "regulated medical waste" term is used for these predefined and categorized medical wastes. Current laboratory guidelines for working with infectious microorganisms at biosafety level (BSL) 3 recommend that all laboratory wastes be decontaminated before disposal by an approved method, preferably within the laboratory *[Center for Disease Control and Prevention, 2003]*. This is even more important considering that the COVID pandemic was a result of improper protection and disposal of laboratory samples.

2.7.1 Contaminants in Clinical Solid Waste

The current updated regulatory body guidelines require the healthcare facility laboratories to be capable of destroying discarded cultures and stocks on-site especially in case these laboratories isolate any microorganism or toxin identified as a "select agent" from a clinical specimen. *Tables 2.5, 2.6,* and *2.7* enlist select agents and toxins from Health and Human Services (HHS) as reproduced from Background I. Regulated Medical Waste Guidelines for Environmental Infection Control in Healthcare Facilities *[Center for Disease Control and Prevention, 2003]*.

Table 2.5: Non-overlap select agents and toxins from Health and Human Services (HHS) [*Center for disease control and prevention, 2003*]

Pathogen type	Select agents
Viruses	Crimean-Congo hemorrhagic fever virus; Ebola viruses; Cercopithecine herpesvirus 1 (herpes B virus); Lassa fever virus; Marburg virus; monkeypox virus; South American hemorrhagic fever viruses (Junin, Machupo, Sabia, Flexal, Guanarito); tick-borne encephalitis complex (flavi) viruses (Central European tick-borne encephalitis, Far Eastern tick-borne encephalitis [Russian spring and summer encephalitis, Kyasnaur Forest disease, Omsk hemorrhagic fever]); variola major virus (smallpox virus); and variola minor virus (alastrim). *Exclusions:* Vaccine strain of Junin virus (*Candid. #1*)
Bacteria	*Rickettsia prowazekii, R. rickettsii, Yersinia pestis*
Fungi	*Coccidioides posadasii*
Toxins	Abrin; conotoxins; diacetoxyscirpenol; ricin; saxitoxin; Shiga-like ribosome inactivating proteins; tetrodotoxin Exclusions: The following toxins (in purified form or in combinations of pure and impure forms) if the aggregate amount under the control of a principal investigator does not, at any time, exceed the amount specified: 100 mg of abrin; 100 mg of conotoxins; 1,000 mg of diacetoxyscirpenol; 100 mg of ricin; 100 mg of saxitoxin; 100 mg of Shiga-like ribosome inactivating proteins; or 100 mg of tetrodotoxin.
Genetic elements, recombinant nucleic acids, and recombinant organisms	Select agent viral nucleic acids (synthetic or naturally-derived, contiguous or fragmented, in host chromosomes or in expression vectors) that can encode infectious and/or replication competent forms of any of the select agent viruses.
	Nucleic acids (synthetic or naturally-derived) that encode for the functional form(s) of any of the toxins if the nucleic acids: • are in a vector or host chromosome; • can be expressed in vivo or in vitro; or • are in a vector or host chromosome and can be expressed in vivo or in vitro; Viruses, bacteria, fungi, and toxins that have been genetically modified.

Table 2.6: High consequence livestock pathogens and toxins/select agents (overlap agents) [*Center for disease control and prevention, 2003*]

Pathogen type	Select agents
Viruses	Eastern equine encephalitis virus; Nipah and Hendra complex viruses; Rift Valley fever virus; Venezuelan equine encephalitis virus Exclusions: MP-12 vaccine strain of Rift Valley fever virus; TC-83 vaccine strain of Venezuelan equine encephalitis virus
Bacteria	*Bacillus anthracis; Brucella abortus, B. melitensis, B. suis; Burkholderia mallei* (formerly Pseudomonas mallei), *B. pseudomallei* (formerly P. pseudomallei); botulinum neurotoxin- producing species of *Clostridium; Coxiella burnetii; Francisella tularensis*
Fungi	*Coccidioides immitis*
Toxins	*Botulinum* neurotoxins; *Clostridium perfringens* epsilon toxin; *Shigatoxin*; staphylococcal enterotoxins; T-2 toxin Exclusions: The following toxins (in purified form or in combinations of pure and impure forms) if the aggregate amount under the control of a principal investigator does not, at any time, exceed the amount specified: 0.5 mg of botulinum neurotoxins; 100 mg of Clostridium perfringens epsilon toxin; 100 mg of Shigatoxin; 5 mg of staphylococcal enterotoxins; or 1,000 mg of T-2 toxin
Genetic elements, recombinant nucleic acids, and recombinant organisms	Select agent viral nuclei acids (synthetic or naturally derived, contiguous or fragmented, in host chromosomes or in expression vectors) that can encode infectious and/or replication competent forms of any of the select agent viruses; Nucleic acids (synthetic or naturally derived) that encode for the functional form(s) of any of the toxins if the nucleic acids: • are in a vector or host chromosome; • can be expressed in vivo or in vitro; or • are in a vector or host chromosome and can be expressed in vivo or in vitro; Viruses, bacteria, fungi, and toxins that have been genetically modified

Table 2.7: Pathogens and endotoxin producers found in the air of waste handling facilities *[Collins and Kennedy, 1992]*

Pathogens	Endotoxin Producers
Acinetobacter calcoaceticus	*Enterobacter agglomerans*
Escherichia coli	*Enterobacter cloacae*
Klebsiella pneumoniae	*Klebsiella oxytoca*
Proteus spp.	*Pseudomonas putida*
Pseudomonas spp.	
Salmonella serotypes	
Serratia marcescens	
Staphylococcus aureus	
Streptococcus pyogenes	
Strebtococcus bneumoniae	

2.7.2 Contaminants in Liquid Medical Waste

Liquid wastes from hospital facilities are usually classified according to their nature and sources of origin under the following groups:

(i) Infectious waste

- Blood and body fluids.
- Laboratory wastes, such as, cultures of infectious agents, cultures from laboratories, biological, discarded vaccines, culture dishes and devices.

(ii) Chemically hazardous

- Formaldehyde (from pathology labs, autopsy, dialysis, embalming)
- Mercury (broken thermometers, sphygmomanometer, dental amalgams)
- Solvents (pathology and embalming)
- Radioactive isotopes (oncology centers)

(iii) Pharmaceutical liquid waste (discarded/unused/expired medicines)

(iv) Photographic chemicals (fixer and developer)

(v) From cleaning and washing water channeled into the drain.

Liquid waste may have a plethora of microorganisms ranging from viruses like HIV, HBV, HSV, vegetative bacteria like *Pseudomonas, Staphylococcus,* and *Salmonella,* fungi (e.g., *Candida*), mycobacterium, such as, *M. tuberculosis* and *M. bovis,* and non-enveloped viruses like *Adenovirus* and *Parvovirus [Biswal, 2013].*

References

- Aggarwal Jyoti, Sharma Swati, Kamyab Hesam, Nadda Ashok, *"The Realm of Biopolymers and their Usage: An Overview,"* J. Environ. Treat Tech., Vol 8, Issue 2, pages 1005-1016, 2020.
- Alafraj Abdullah A, Lee Henry Hsin-chung, Munusamy Murugan A, Ling Qing-Dong, Kumar Suresh, Chang Yung, Chen Yen-Ming, Lin Hong-Ren, Lu Yi-Tung, Wu Gwo-Jang, Higuchi Akon, *"Development of biomaterial surfaces with and without microbial nanosegments,"* J Polym Engg; 36(1): pages 1–12, 2016.
- Biswal Sasmita, *"Liquid Biomedical Waste Management: An Emerging Concern for Physicians,"* MJMSR, Vol- 4, Issue - 2, pages 99-106, 2013.
- *Centre for Disease Control and Prevention,* Background I. Regulated Medical Waste Guidelines for Environmental Infection Control in Health-Care Facilities, 2003.
- Collins C.H. and Kennedy D.A., *"The Microbiological Hazards of Municipal and Clinical Wastes,"* Journal of Applied Bacteriology, 73, pages 1-6, 1992.
- Duvetter T., Sila D.N., Van Buggenhout S., Jolie R., Van Loey A., Hendrickx M., *"Pectins in Processed Fruit and Vegetables: Part I—Stability and Catalytic Activity of Pectinases,"* Comprehensive Rev in Food Science and Food Safety, Vol - 8, Issue - 2, 2009.
- M. Ebara, et al., *Smart Biomaterials,* Springer, Tokyo, Japan, 2014.
- Environment (Protection) Act, 1986, Gibbs, Mo, and Steele, P., *"Post-Harvest Technology of Horticultural Crops,"* ED-Tech Press, 2020.
- Gill Alexander, *"The Importance of Bacterial Culture to Food Microbiology in the Age of Genomics,"* Front Microbiol; 8: pp. 777, 2017.
- Grand View Research, Market Analysis Report, Biomaterials Market Size, Share & Trends Analysis Report by Product (Natural, Metallic, Polymer), by Application (Cardiovascular, Orthopedics, Plastic Surgery), by Region, and Segment Forecasts, 2020 – 2027, Report ID: GVR-2-68038-637-0, Feb 2020.
- Henson Spencer and Traill William, *"The demand for food safety: Market imperfections and the role of government,"* Food Policy, Vol. 18, Issue 2, pp. 152-162, 1993.
- Hueston Will and McLeod Anni, *"Overview of the Global Food System: Changes Over Time/Space and Lessons for Future Food Safety,"* In: Institute of Medicine (US). Improving Food Safety through One Health Approach: Workshop Summary. Washington (DC): National Academies Press (US). A5. Available from: https://www.ncbi.nlm.nih.gov/books/NBK114491/, 2012.
- Ige Oladeji O., Umoru Lasisi E., Aribo Sunday, *"Natural Products: A Minefield of Biomaterials,"* ISRN Materials Science, Vol 2012, pp. 1-20, 2012.
- Jon, S.' Biomedical Nanomaterials, Vol. 13, (2018).
- Kohli P, Kalia M, Gupta R, *"Pectin Methylesterases: A Review"* J Bioprocess Biotech 5: 227 DOI:10.4172/2155-9821.1000227, 2015.

- Langeveld, J.P.M, Wang, J.J, van de Wiel, D.F.M, Shih, G.C, Garssen, G. J, Bossers, A. & Shih, J. C. H; *"Enzymatic Degradation of Prion Protein in Brain Stem from Infected Cattle and Sheep"*, The Journal of Infectious Diseases, 188(11), pp.1782-1789, 2003.
- Lim, J.O; Huh, J.S; Abdi, S.I.H; Ng, S.M; Yoo, J.J; *"Functionalized Biomaterials-Oxygen Releasing Scaffolds"*, Biotechnol Biomater, 5 (2), 2015.
- Markets and Markets, *"Plant Extracts Market by Type (Phytomedicines & Herbal Extracts, Essential Oils, Spices, Flavors & Fragrances)"*, Application (Pharmaceuticals & Dietary Supplements, Food & Beverages, Cosmetics), Source, and Region - Global Forecast to 2025, REPORT CODEFB 1399, Jul 2019.
- Martinez V.M, and Whitaker J.R, *"The Biochemistry and Control of Enzymic Browning,"* Trends. Food. Sci., 6, pp.195, 1995.
- Mishra Bibhuti Bhusan and Gautam Satyendra, *"Polyphenol Oxidases: Biochemical and Molecular Characterization, Distribution, Role and its Control"* Enz Eng, 5:1, 2016.
- Nagodawithana T. and Reed J.R, *Enzymes in Food Processing,* Academic Press Inc., California, 1993.
- National Institutes of Health, National Human Genome Research Institute, Talking Glossary of Genetic Terms: Protein. https://www.genome.gov/glossary/, 2019.
- Nooreen Zulfa, Rai Vineet Kumar, Yadav Narayan Parasad , *"Phyto-Pharmaceuticals: A New Class of Drug in India,"* Annals of Phytomedicine 7(1): 27-37, 2018.
- Prigge S.T, Boyington J.C., Faig M., Doctor K.S., Gaffeney B.J., Amzel L.M, *"Structure and Mechanism of Lipoxygenases,"* Biochimie, 79, pp.629-636, 1997.
- Rimondini L., Fini M., Giardino R., *"The Microbial Infection of Biomaterials: A Challenge for Clinicians and Researchers, A Short Review,"* Journal of Applied Biomaterials & Biomechanics; Vol. 3 No. 1, pp.1-10, 2005.
- Saini Savita, Das Bimal K, Kapil Arti, Nagarajan Shyama S, Sarma R.K, *"The Study of Bacterial Flora of different types in Hospital Waste: Evaluation of Waste Treatment at AIIMS Hospital, New Delhi,"* Southeast Asian J. Trop Med. Public Health, Dec; 35(4) pp. 986-989, 2004.
- Silva Filipa VM, Sulaiman Alifdalino, *"Chapter 4: Advances in Thermosonication for the Inactivation of Endogenous Enzymes in Foods,"* Ultrasound: Advances for Food Processing and Preservation, pp.101-130, 2017.
- Soares, Goncalo C.; Learmonth, David A.; Vallejo, Marian C.; Perez Davila, Sara; Gonzalez, Pio; Sousa, Rui A. ; Oliveira, Ana L. ; *"Review: Supercritical CO_2 Technology: The Next Standard Sterilization Technique"*, Materials Science & Engineering C. 99, pp.520–540, (2019).
- Tomasko, D.L. et al., *"A review of CO_2 Applications in the Processing of Polymers"*, Ind. Eng. Chem. Res. 42 (25) pp. 6431–6456, (2003).

- US FDA , *"Water Activity (aw) in Foods, Inspection",* Technical Guides, 2014. (https://www.fda.gov/inspections-compliance-enforcement-and-criminal-investigations/ inspection-technical-guides/water-activity-aw).
- Vaisanen, T.; Das, O.; Tomppo, L.; *"A review on New Bio-based Constituents for Natural Fiber-Polymer Composites",* J. Clean. Prod. 149 pp. 582–596, 2017.
- Voragen V.B, *Biotechnological Innovations in Food Processing,* 3rd Edition, Butterworth-Heinmann, New York, 1991.
- Zhang, J. ; Davis, T.A. ; Matthews, M.A.; Drews, M.J.; LaBerge, M. ; An, Y.H.; *"Sterilization using High-Pressure Carbon Dioxide",* J. Supercrit. Fluids 38 (3), 2006.

❐

3

Characterization Methods and Evaluation of Sterility

3.1 Relevance of Microbial analysis in Food Industry

Food industry is a sensitive industry, as it directly impacts the health of a consumer. It thus becomes imperative to follow safety measures in order to ensure that the food is free from harmful microbes. Preventive measures, such as, cleaning and sanitation, temperature control of the production process, hygiene of the site and workers, safe handling and transportation etc., are essential pre-requisites to avoid microbial contamination. Emphasis is given to the reduction or elimination of contaminants during the production process such that it does not adversely affect the quality of the food product. Accordingly, the food safety management system requires microbial analysis of foods in order to ensure that products and processes are within the set specifications.

It is important to note that inactivation, recontamination, and growth can occur at several steps of the production process. If the inactivation process eliminates microorganisms and if subsequent recontamination is avoided then the production is under control. In case there is presence of low counts of microorganisms in the final product, arresting their growth by process control and packaging can ensure its safety for consumption. There are standard testing protocols for microbial analysis that have been set by the regulatory bodies, and are followed for a food production process, as can be seen from the flow chart *(Figure 3.1)*. Good safety management should be based on evidence

M. Mukhopadhayay and A. Chatterjee, *Sterilization and Preservation*,
https://doi.org/10.1007/978-3-031-17370-7_3

that hazards are well under control and that the final level of the hazard is within acceptable limits irrespective of initial levels of organisms, reduction, recontamination, and growth.

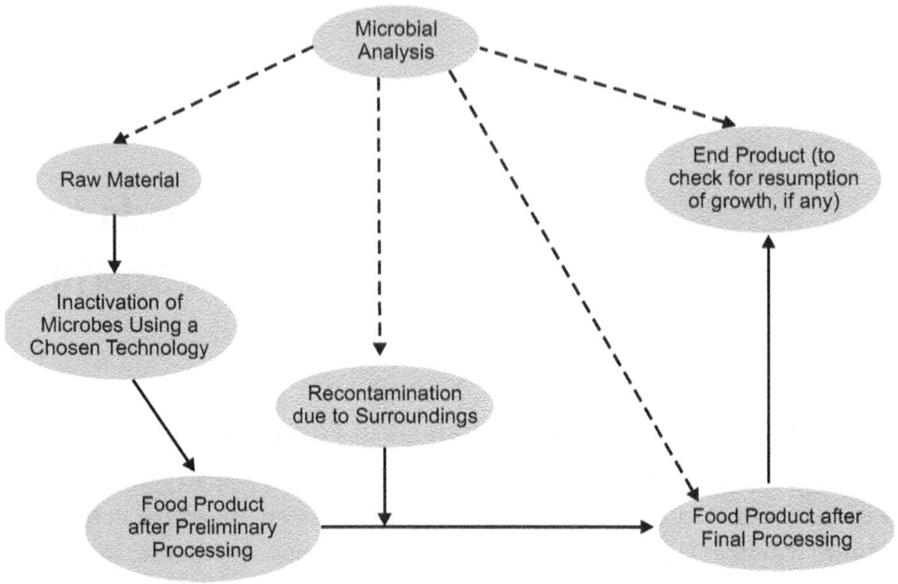

Fig. 3.1: Microbial analysis protocol in a food production process. *[Adapted from Zwietering et al., 2016]*

3.1.1 Sources of Contamination

Contamination can occur at any point in the food system. The scale and complexities of today's food systems (e.g., packaged food, precooked food, 'ready-to eat' food, fruit juices, fruit and spice pastes, etc.) contribute to the likelihood and magnitude of food-borne pathogens. The more complex it is, the more opportunities it creates for contamination; the larger is the reach, the more people will be potentially affected. There are several factors, including physical and/or chemical treatment of food during processing, that cause injury to foodborne pathogens. The injury can be a sub-lethal injury, which is either permanent or temporary, resulting in the microbe remaining undetected in the food system. This is of serious concern, as harmful enzymes within the microbial cell may leach out in food system contaminating it.

3.1.2 Standard Testing Methods for Microbes

The commonly used norms of microbial analysis in food systems are detection and enumeration. The presence of specific bacteria and their concentration need to be determined, to carry out further assessment and subsequent control of safety hazards. This also enables to check the potential for spoilage and retain the product characteristics. It has been a constant endeavour of scientists working in the research field of food safety and quality, to search for new methods of detection of microbes that are sensitive, accurate, rapid, economical, and effective in determining potential microbial risks. Conventional cultivation methods and phenotypical tests are till date the most used in the food microbiological field, in spite of being time-consuming and laborious. In fact, the conventional method is still considered as the 'gold standard'. The past decade has seen a tremendous advancement in this field with great advancement and improvisations in the tools for molecular diagnosis. Immunoassays, mass spectrometry, and PCR-based technologies have also been recently introduced for microbial identification.

In order to determine subtle and specific differences in the bacterial genome, and to determine the accurate bacterial species, microarrays are being used. These are powerful tools owing to high sensitivity, and specific high throughput that enables accuracy in identification based on single target detection. Similarly, there is an enormous potential for the detection and identification of bacterial species using gene arrays, as they can hybridize multiple DNA targets simultaneously, thus speeding the identification process. Likewise, reverse dot-blot (RDB) hybridization coupled with membrane-based macroarrays is comparatively new, yet a cost-effective alternative that can be used for effective and specific identification of food borne microbes. The microflora in food products can differ significantly in their pathogenicity and how it affects spoilage, thus there is a greater need to detect specifically those that represent a high microbial risk *[Böhme et al., 2014]*.

Food borne pathogens are detected using various techniques, which can be broadly classified as culture-based and culture-independent (nucleic acid–based and phage-based) methods.

3.1.2.1 Culture-based Methods

Culture-based methods are long known and traditionally used for microbial food analysis, widely regarded as the 'gold standard'. This method is based on the ability of the bacteria to grow and form visible colonies, when plated on laboratory culture media. The culture based methods are sensitive, economical, convenient, and give either qualitative or quantitative information on the number and type of viable microorganisms present in the samples. However, this method is laborious and time-consuming as the incubation period varies between 18-72 hours. Also, there is a good chance of error as only those cells can be counted which are cultivable under control conditions (e.g., incubation temperature, incubation time, selective media and oxygen availability), while the damaged yet living cells or the cells with sub-lethal injury may resume growth while remaining undetected.

i) **Preparation of Culture and Growth Medium:** The stock culture of yeast is maintained at 4°C on agar slants containing 2% yeast extract, 1% peptone, 2% dextrose and 2% agar-agar (HIMEDIA, India). The growth medium consists of 2% yeast extract, 1% peptone, and 2% dextrose in 100ml distilled water. A loop full of the culture is inoculated in this medium and grown at 30°C, till a concentration of 10^7CFU/ml is reached. The stock culture of *Escheria coli, Bacillus subtilis* and *Lactobacillus rhamnosus* are maintained on Luria-Bertanii (LB) agar slants and *Lactobacillus* MRS agar slants (HIMEDIA, India) respectively at 4°C. A loop full of the cultures are inoculated in LB broth and grown at 37°C, till a concentration of 10^7CFU/ml is reached.

   To obtain the growth of a mixed population, a loop full of Yeast, *E. coli* and *Lactobacillus* are inoculated in LB broth and 2% dextrose is added to the medium to facilitate the growth of yeast. The cultures are grown at 37°C till a population of 10^7CFU/ml is reached. The samples are serially diluted using sterile saline solutions and plated on YPD agar for yeast, LMRS for *Bacillus* and *Lactobacillus* and Nutrient Agar (NA) for *E. coli.* In case of mixed population, 1 ml of the diluted culture is plated on NA, LMRS as well as YPD, using spread plate technique and the colonies are counted. Samples are plated before treatment and also after treatment. The medium is sterilized and poured in sterile glass petri plates and incubated at 37°C for 24 hours to check for contamination. The media and glass wares are sterilized in an autoclave at 0.2MPa, 120°C for 15minutes *[Chakraborty, 2006].*

ii) **Cell Concentration Measurement:** The cell concentrations are measured using a UV-Visible spectrophotometer (Jasco V-530, Japan) at 600nm. The broth is centrifuged for 15minutes at 10,000rpm (after suitable dilutions) in order to separate the cell mass. The separated cell mass is resuspended in 0.9% NaCl solution. The cell optical density is read against a blank containing 0.9% NaCl. (An absorbance reading 1 corresponds to a population of 10^7 cells). The growth curves are plotted as shown in *Figure 3.2* and *Figure 3.3* for *E. coli* and *Sacharomyces cerevisiae* and *Sacharomyces cerevisiae* respectively, wherein O.D represents the optical density as observed in an UV spectrophotometer *[Chakraborty, 2006].*

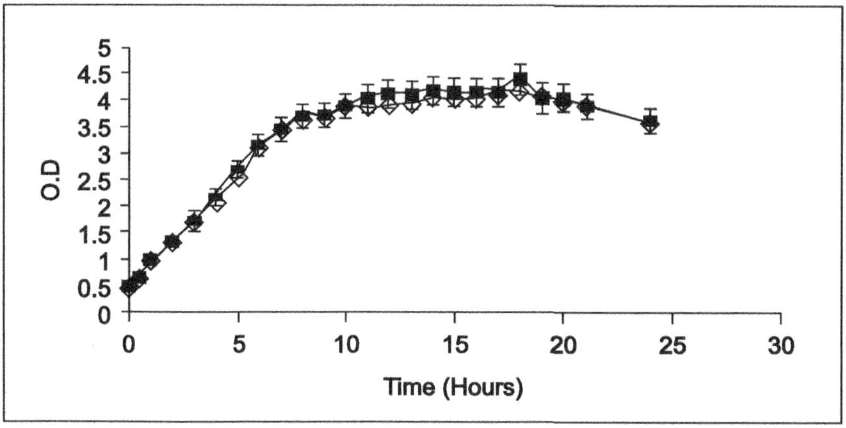

Fig. 3.2: Growth curve for *Escherichia coli [Chakraborty, 2006]*

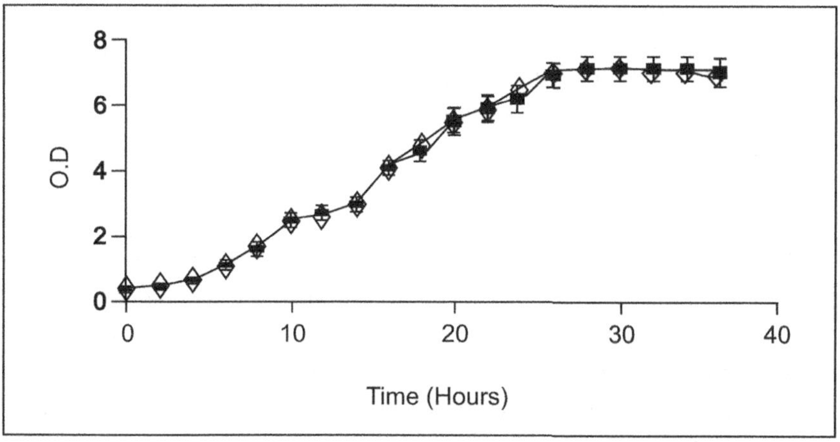

Fig. 3.3: Growth curve for *Sacharomyces cerevisiae [Chakraborty, 2006]*

iii) Determination of Microbial Count and Shelf Life: Microbial count in a food system is the indicator of safety, shelf life, and the extent of sterilization achieved. The Food and Drug Administration body of each country has certain specific rules and regulations regarding the maximum allowable number of microorganisms in a particular food system. To check the efficiency of sterilization, the untreated and treated samples are plated and the microbial colonies counted. The plate count method of estimating the viable population of a sample is used to check the microbial count. The samples after the serial dilution are prepared with sterile saline. The dilutions are mixed thoroughly to ensure that bacterial clumps are dispersed evenly. Separate sterile pipette tips are used for each dilution. A 1 ml sample of each dilution is then plated on the agar plates using the spread plate technique. The dilutions of the samples are plated on NA, LMRS and YPD. The plates are then incubated for 24 to 48 hours. A plate containing more than 300 colonies should not be selected. A plate with countable colonies is selected and the number of colonies present is counted. From this count the viable population of the original sample can be calculated. The shelf life is ascertained by the standard procedure of keeping the treated samples in a series of sterilized glass tubes plugged with sterile non-absorbent cotton (without refrigeration). Samples from the test tubes are serially diluted and plated on NA, LMRS and YPD. The plates are incubated for 24 to 48 hrs and the shelf life is obtained when the growth in the sample reached the maximum allowable limit set by the US Food and Drug Administration.

3.1.2.2 Culture-independent Methods

There are several culture independent methods to detect the presence of bacteria in food, which include Biochemical detection techniques, Biosensor based detection, Spectroscopic detection techniques, and Instrumental techniques for detection of microflora. However, there are two culture-independent approaches that may provide promising alternatives to culture-based approaches in food industry, namely, nucleic acid–based and bacteriophage-based detection methods.

Table 3.1 summarizes advantages and limitations of culture-independent nucleic acid and phage-based approaches for microbial analysis *[Foddai and Grant, 2020]*.

Table 3.1: Comparison of Nucleic acid–based and Bacteriophage-based detection methods.

Type of test	Test name	Advantages	Limitations
Nucleic acid-based	Reverse-transcriptase qPCR (RT-qPCR)	Quick compared to culture, but longer than viability PCR/qPCR	False positive results may occur RT-qPCR viability assessment is validated for longer (>200 bp) transcription products, but not necessarily for short qPCR products
	Viability PCR/ qPCR	Differentiate between viable and dead cells quickly compared to culture	Inactivated bacterial cells do not always have compromised cell membranes, leading to false positives
Phage based	Phage amplification + qPCR	Rapid test to detect released phages or the host DNA by qPCR, which demonstrates that lysis has occurred It is a quantitative assay	Important to control the DNA release to Maximise detection senstivity
	Phage amplification (Plaque) assay	A 24-h test, producing countable plaques A quantitative result	Not a high- throughput test Laborious multi-step test
	Phage amplification + immunoassay	Rapid test Only viable cells lyse hence potentially a quantitative assay	Analytical senstivity limited
	Phage amplification + enzyme assay	Rapid test It is a qunatitative assay	Genetically engineered phage may be required

3.1.3 Environment Scanning Electron Microscope (ESEM)

Scanning electron microscope (SEM) is the microscope which uses the electrons instead of light to magnify the image of the specimen by several times than that of an ordinary optical microscope. The beam of the electrons is focused on the specimen through the magnetic lenses. The illuminated sample (which is made conductive by coating with gold or carbon) emits the secondary electrons from its surface. These secondary electrons when captured produce the enlarged image of the specimen.

The microbial cell structures and morphology are examined by ESEM to understand the effect of physical stresses exerted on the cell wall and in the cell internals. This method of characterization is utilized to establish the mechanism of inactivation of microorganisms by the sterilization process involved. *Figure 3.4* gives the protocol for sample preparation.

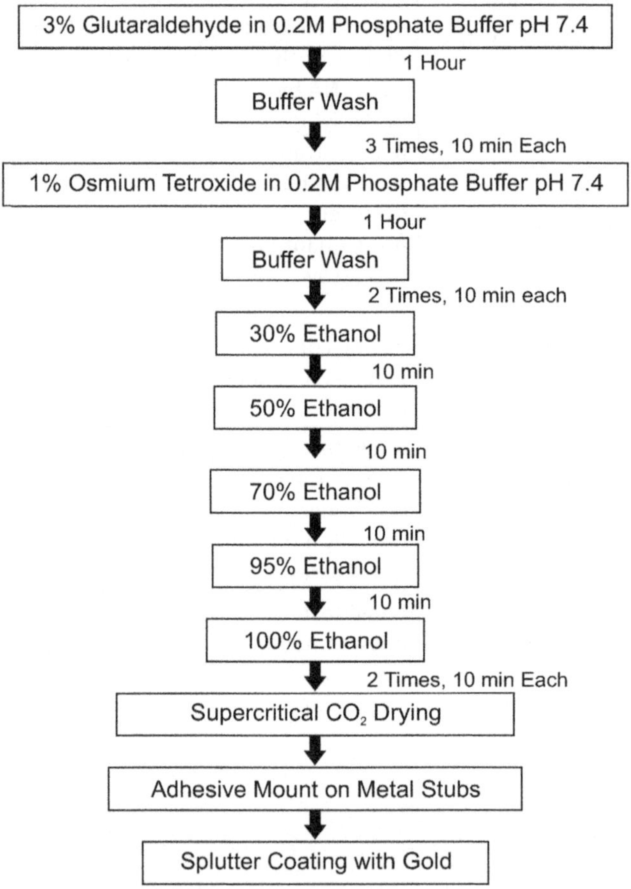

Fig. 3.4: Protocol for soft tissue sample preparation for SEM.

3.1.4 Limits of Detection and Requirements of Inactivation

The limit of detection (LOD) for the various methods of analysis for microflora is the minimum concentration of cells that can be detected. Examining the presence of bacteria causing foodborne intoxication or spoilage does not generally require limits of detection below 100 CFU/g or ml. Microbes causing spoilage can impact the quality of a product only when they exceed a significant concentration, for example spoilage of red meats by *Pseudomonads* becomes apparent above 6 log CFU/cm^2 *[Gill, 2017]*. Similarly, for *Bacillus cereus* populations of > 10^6 CFU/g are needed to produce toxin at levels hazardous to health. For *Clostridium perfringens,* a population of >10^5 CFU/g is needed to result in illness; therefore, a 3 log increase would control the hazard, *E. coli*. The inherent variability in quantitative methods necessitates the use of a progressive increase of < 1 log as an indicative that growth is controlled. However, 1 log increase is concluded as an appropriate level of control for *L. monocytogenes*, whereas in the case of *Staphylococcus aureus*, no detectable toxin may be formed under certain sterilizing condition. As with *C. botulinum*, the current methodology should be used for toxin detection and specific toxin levels should be determined. In lieu of testing for toxin, limiting growth to < 3 log may be used. This limiting growth level is based on an initial population of 1000 CFU/g and a minimum of 10^6 CFU/g to produce toxin *[FDA, 2001]*.

3.1.5 Standard Testing Methods for Enzyme Activity

It is a standard procedure to conduct a risk assessment of enzymes in foods. Some of the enzymes have negative effect on the food product in terms of quality, texture, taste, and hence it is essential to denature or inactivate them. On the other hand, there are certain enzymes, which are the main bioactive molecules which make the said food a 'super food'. Thus, it is necessary to assess the enzyme activity in the food at the end of the process for judicious selection of the method of sterilization and stabilization of food products.

3.1.5.1 Color Measurement

The color of food is an indicator of the freshness of the food and the deterioration that might have occurred during processing. The enzymes and microorganisms bring about a change in the food color if they are not inactivated or incompletely inactivated. On the other hand, if a process makes use of very high temperatures it will cause non-enzymic browning and charring which, in turn, will affect the original color of the food. A color change may not necessarily change the nutritive value of a food. However, it does affect the consumers' acceptance of such foods. Color measurement is a critical objective quality parameter that can be used for the following applications: (i) quality index measurements of raw and processed foods, (ii) quality control documentation and communication, (iii) determination of conformity of food quality to specifications, and (iv) analyses of quality changes as a result of food processing, storage, and other factors *[Giese, 2000]*.

The Hunter's Color Lab instrument is usually used to measure the color change in the untreated and treated samples. The instrument measures the degree of lightness (L), the degree of redness (+a) or greenness (-a), and the degree of yellowness (+b) or blueness (-b). Lightness is expressed by "L" values, and "a" and "b" values are the chromaticity coordinates. Lightness is expressed as dark to light color with 0 = black and 100 = white. Green to red is expressed by the "a" value, the more positive value being represented more red, while the more negative value being represented more green color (-80 = green, 100 = red). Blue to yellow is expressed by the "b" value; whereas a more positive value represents more yellow color in the sample (-80 = blue, 70 = yellow). The colorimeter is calibrated with a standard white tile. A 5ml sample is placed in the glass cup and covered with the standard white tile. On comparison of the treated samples with the fresh samples, the degree of color deterioration or color change is determined, which in turn, determines the harshness or efficiency of the process.

3.1.5.2 H-NMR Analysis of Polysaccharides

This test is performed for juices (e.g., *Aloe vera* juice) to quantify mannose and polysaccharide, the main bioactive substances present in *Aloe vera*. The treated, untreated and commercial samples of *Aloe vera* juice are dissolved in 0.5ml of heavy water (D_2O). The H-NMR (proton Nuclear Magnetic Resonance) spectra are recorded on a 500 MHz Varian spectrometer, with single 90° pulse at a temperature of 45°C. All H-NMR (Varian, Mercury Plus) spectra are taken with a sweep width of 8000Hz. The residual HOD resonance is pre-saturated during the second delay period. Dioxane is used as an external standard (67.5ppm).

3.1.5.3 Lipoxygenase Assay

The spectrophotometric method is used to carry out the *lipoxygenase* assay in the treated and untreated samples of soybean and rice bran. *Lipoxygenase* causes rancidity in rice bran; thus to prevent off-flavors and increase shelf life, it must be inactivated. The following method can be used to quantify the deactivation of the enzyme. The reaction mixture containing 0.5ml of 200mM sodium borate buffer of pH 9, 0.1ml of 100μM methylene blue, and 0.1ml of 10mM sodium linoleate substrate are mixed with 0-0.3ml of soybean extract sample and distilled water to make the final volume to 1.0 ml. The reaction is initiated by the addition of the sample and the absorbance at 660nm is recorded with a spectrophotometer (Jasco V-530, Japan) for 5 min at 25°C. The test is performed as follows: 2.5mg of soybean flour is weighed into a test tube, 0.5ml of distilled water is added to it and the mixture is allowed to stand for 5 to10 min. Dye-substrate is prepared by mixing 25ml of 200mM sodium borate buffer, 5ml of 100μM methylene blue, and 5ml of distilled water in a glass stoppered bottle. 2ml of this dye-substrate is added to the sample. After 3min the color of the sample is checked against the dye-substrate as the blank. The extent of enzyme deactivation is determined by comparison with the raw samples *[Suda et al., 1995]*.

3.1.5.4 Trypsin Inhibitor Assay

Trypsin inhibitor (TI) is categorized as ANF (anti nutritional factor), which interferes with digestion activity and has an antinutritional effect. This enzyme needs to be deactivated to render a food beneficial for health. The method described can be used to check the extent of denaturation of the enzyme. The samples of soybeans are diluted using distilled water. To each of the four test tubes 2ml of this diluted sample is added. A 5th test tube is prepared for the trypsin standard by adding 2ml of distilled water. To three of the remaining four test tubes containing the sample and the 5th test tube, 2ml of trypsin solution is added and kept at 37°C for 10min. After incubation, 5ml of BAPA (benzoyl-DL-arginine-*p*-nitroaniline) solution is added to all the test tubes and stirred immediately on a vortex. This is incubated again at 37°C for 10 min, and the reaction is terminated by adding 30% acetic acid. A sample blank is prepared except that the trypsin is added after the termination of the reaction. Absorbance is determined at 410nm against the sample blank using a spectrophotometer (Jasco V-530, Japan). The trypsin inhibitor activity is calculated as *[Hamstrand, 1981]:*

$$TI\ mg/g\ of\ sample = \frac{\text{Standard} - \text{Sample}}{0.019 \times 1000} \times \text{Dilution Factor}$$

3.1.5.5 Free Fatty Acid Analysis

Free fatty acid (FFA) is primarily responsible for the off-flavors and rancidity in oils. To ensure that the oil has prolonged shelf life, the activity of FFA needs to be arrested. The method described here can be used to determine the FFA activity. 50ml alcohol is taken in a clean dry 150ml flask and a few drops of the oil or the sample and 2ml phenolphthalein are added to it. The flask is placed in a hot water bath at 60-67°C until it becomes warm. To this sufficient 0.1M NaOH is added to produce a faint permanent pink color. Then 56.4 g oil or sample is added to the neutralized alcohol and titrated with 0.1M NaOH, while occasionally warming and vigorously shaking the mixture until the same faint pink color appeared in the supernatant alcohol *[AOAC, 2000]*. The percent free fatty acid (expressed as oleic acid) is reported as:

$$\% FFA = Volume\ ml\ of\ 0.1M\ NaOH \times 0.05$$

3.2 Relevance of Microbial analysis in Biomaterials

The safety and successful application of biomaterials and medical devices require their microbial analysis at different stages in order to ensure that sterilized products and processes are within the set specifications.

3.2.1 Factors for Contamination

The causes of contamination of medical devices arise from many peculiar factors, including the chemistry of the biomaterial, the physical properties of the surface, the design of the medical device, the anatomical site, the extent of surgical invasion, and the time of application. In addition, the environmental conditions and response of the host also contribute to the development of colonization and infection *[Rimondini et al., 2005]*.

3.2.2 Detection and Testing Methods

Most common bacterial and fungi spp. involved in biomaterial infections belong to endogenous bacteria, such as, *Staphylococcus epidermidis* and *Staphylococcus aureus*. The consequences of microbial infections arising from the clinical applications of biomaterials may be varied, as can be seen from biological indications listed in *Table 3.2 [Rimondini et al., 2005]*.

Table 3.2: Biological indications and clinical inference

Biological Indication	Clinical Inference
Inflammation with abscess	Infection
Sequestrum presence	Infection near abscess
Brain abscess	Respiratory and systematic infection
Inflammation with fibrosis	Infection confirmed only if there is bacterial presence
Intracellular bacteria	Infection

A microbial contamination of the medical device will eventually develop into a biofilm, which is the survival tactics of the microbial colony in the hostile environment. In the form of a biofilm the microbial colonies optimize their nutrient uptake and protect each other from the removal forces of the host defence mechanism and the toxic action of an antibiotic. An effective method to detect presence of microbial biofilm is the triphenyl tetrazolium chloride (TTC) assay. In this assay, soluble colorless TTC is reduced to insoluble red formazan by electron transfer that is associated with active oxidative bacterial metabolism, which is precipitated intracellularly. TTC, as a metabolic indicator for presence of bacterial biofilm, requires minimal setup time *[Rózalska et al., 1998]*.

To identify and classify bacteria in taxonomical division, gram staining is used, categorizing the bacteria in two major groups – gram-positive and gram-negative. The gram-positive bacteria (gram+) is one which has a simple, thick and relatively impermeable wall structure that is composed of peptidoglycans and secondary polymers, whereas, the gram-negative bacteria (gram−), has a complex cell wall structure that is composed of thinner peptidoglycan layers and a lipid-protein bilayer covering it, known as the outer membrane.

The following tests are followed to validate and establish a sterilization procedure,

i) The change in morphological structure of the material.
ii) The effect on the bioactive component.
iii) The structure and mechanical properties.
iv) Efficiency of sterilization in comparison to other methodologies.
v) Physicochemical characterization of the materials before and after sterilization.
vi) The incubation time, which is commonly 14 days as per standards, such as, ISO 11737-2:2009 for medical devices.

Techniques, such as, Fourier Transform Infrared Spectroscopy, SEM, tensile testing, tear testing, etc., are used to characterize different synthetic and natural materials before and after the sterilization treatment. In addition, their structure, morphology, pH, water content, and viscoelastic properties are measured *[Karajanagi et al., 2011]*. Moreover, biomaterials are tested to assess if they are biocompatible and non-toxic when implanted subcutaneously in ferrets. It is possible to have catheters sterilized without enhancing the binding capacity of microorganisms on the surface of the catheter shell, as some sterilization process may induce a modification on the surface of the catheter shell, which could be responsible for enhancing the binding capacity of microbial cells *[Bertoloni et al., 2009]*.

3.2.3 Standards of Sterility and Sterilization Efficiency

Sterility may be defined as the absence of viable microorganisms including viruses from the sterilised product. Testing for sterility involves immersing or flushing the component with sterile microbial growth medium, incubation of the medium under conditions favorable for microbial growth, and observation of turbidity or other indication of microbial growth after a suitable incubation period. The destruction of microorganisms by physical or chemical agents follows an exponential law. Accordingly, one can calculate a finite probability of a surviving organism regardless of the magnitude of the delivered sterilization dose or treatment. The probability of survival is a function of the number and types (species) of microorganisms present on the product (bioburden). A sterility assurance level (SAL) is derived mathematically and it defines the probability of a viable microorganism being present on an individual product unit after sterilization. SAL is normally expressed as 10^{-n}. Estimation of the minimum required dose of sterilant for sterilization (to achieve a SAL 10^{-6}) and for decontamination (reducing 5 to 6 orders of magnitude) is established using as reference, a representative virus family member. A terminally sterilized unit is generally accepted to be sterile when it attains a sterility assurance level of $\leq 10^{-6}$. A sterility assurance level (SAL) is the probability of occurrence of a single viable microorganism in a product post sterilization and this count must be less than one in a million (10^{-6}) for a product to be classified as sterile (SAL-6).

For radiation sterilization of microorganisms other than viruses, the ISO 11137 (part 1 and 2) are followed. The radiation resistance of a microorganism is measured by the so called decimal reduction dose (D10 value), which is defined as the radiation dose (kGy) required to reduce the number of that microorganism by 10-fold (one log cycle) or required to kill 90% of the total number. The D10 value can be measured graphically from the survival curve; the slope of the curve (mostly a straight line on logarithmic scale) is related to the D10 value.

National regulatory bodies of each country are responsible for setting the standards which assure the safety, efficacy, quality, and reliability of the drugs and medical devices, right from their developmental stage to the final end product. However, the Food and Drug Administration (FDA, US), the European Medicines Agency (EMA, EU), the International Organization for Standardization (ISO), and the Pharmaceuticals and Medical Devices Agency (PMDA, Japan) are widely accepted as the standard references. Clean rooms having controlled supply, distribution, sterile environment, and air filtration are used during the production of medicines. This is done in keeping with ISO 14644-1:2015, [*2015*] standards, to limit the particle and microbial contamination to acceptable levels. Some pharmaceutical formulations, mainly for parenteral and ophthalmic administration, need to guarantee a sterility assurance level of 10^{-6} (SAL-6), thus requiring an additional sterilization method to achieve the legal requirements *[Ribeiro et al., 2019]*.

Industry standards for validation of sterilization of healthcare products by irradiation have been established by organizations, such as, the American National Standards Institute (ANSI), the Association for the Advancement of Medical Instrumentation (AAMI), the International Organization for Standardization (ISO), and ASTM International (formerly the American Society for Testing and Materials). Advancements in the biopharmaceutical and biotechnology drug industry have led to the need for pre-sterilized or microbially controlled products that can be directly incorporated in critical manufacturing processes. Initially, these disposable products were targeted for small-scale applications, such as, laboratory-scale drug development and preclinical studies.

Several industry standards are used for sterilization validation of gamma irradiated healthcare products. The ANSI/AAMI/ISO 11137:2006 standard (*Sterilization of Health Care Products — Radiation)* was published originally in 1994 (ISO 11137:1994); the document was subsequently revised in 2006 and divided into three parts. These documents provide guidelines and offer strategies for validations of a sterilization process that render a product free from viable microorganisms. At present, a sterility assurance level (SAL) of 10^{-6} is generally accepted for pharmacopoeial sterilization procedures. The attainability of this condition is doubtful, at least in case of non-thermal conventional procedures that are used for sterilization. An additional terminal sterilization step may be needed to attain such stringent requirement *[von Woedtke and Kramer, 2008]*.

Owing to their high resistance, spores are considered as the standard to test the SAL of a sterilization process. The most frequently used spores for this purpose are *G. stearothermophilus,* (a model used to validate steam and hydrogen peroxide sterilization), *B. atrophaeus,* (a model used to validate dry heat and ETO sterilization), and *B. pumilus (* a model used to validate radiation sterilization *[Zhang et al., 2006]*.

Over the last few decades, the success of sterilization of several types of microflora has been indicated in terms of sterilization efficiency and complete inactivation of microorganisms according to the Sterility Assurance Level (SAL). The standards of SAL imposed are enumerated as EN 556-1 and ISO 14937-2009 standards *[Spilimbergo and Bertucco, 2003; Zhang et al., 2006]*.

3.3 Standards of Sterility Levels for Clinical Waste Management

The clinical waste management program involves sterilization of CSW for safe disposal into the environment. The sterility levels to be attained by any treatment technology adopted for inactivation of the infectious pathogens should conform to the standard regulatory norms. *Table 3.3* lists the requirement of levels of sterility to be attained after microbial inactivation by any treatment technology for clinical solid wastes which should have the ability to inactive the biological indicators by 6 log reduction with initial concentration of 10^6 cells/mL. This is as per the guidelines for medical waste treatment technology report on the State and Territorial Association on Alternative Treatment Technologies (STAATT) : 2005 [STAATT, 2005].

Table 3.3: Reduction level requirements of microbial load in CSW as per STAATT : 2005 Regulations

Reduction Level	Requirement
Level I	6 log inactivation of vegetative bacteria, fungi and lipophilic viruses.
Level II	6 log inactivation of mycobacteria.
Level III	4 log inactivation of *B. stearothermophilus* or *B. subtilis* spores.
Level IV	6 log inactivation of *B. stearothermophilus* spores.

References:

- AOAC, The Association of Official Analytical Chemists, Official Methods of Analysis, 17th Edition, Gaithersburg, MD, USA. Methods 925.10, 65.17, 974.24, 992.16, 2000.
- Bertoloni Giulio, Rassu Mario, Vezzù Keti, *"Dense CO_2: An Innovative Method for Sterilization of Venous Catheters at Low Temperature,"* Proceedings 9th International Symposium Supercritical Fluids, 2009.
- Böhme K., Cremonesi P., Severgnini M., Villa Tomás G., Fernández-No I. C., Barros-Velázquez J., Castiglioni B., Calo-Mata P., *"Detection of Food Spoilage and Pathogenic Bacteria Based on Ligation Detection Reaction Coupled to Flow-Through Hybridization on Membranes,"* BioMed Research International, Article ID 156323, 11 pages, 2014.
- Chakraborty, Anuradha, *"Sterilization and Stabilization of Food Products Using Supercritical Carbon Dioxide",* Ph.D. Dissertation, Indian Institute of Technology Bombay, 2006.
- FDA, Comprehensive Reviews in Food Science and Food safety, *"A Report of the Institute of Food Technologists for the Food and Drug Administration of the United States Department of Health and Human Services",* IFT/FDA Contract No. 223-98-2333 Task Order No. 4, Dec 2001. https://www.fda.gov/

files/food/published/Evaluation-and-Definition-of-Potentially-Hazardous-Foods.pdf.

- Foddai Antonio C.G, Grant Irene R, *"Methods for Detection of Viable Foodborne Pathogens: Current state-of-art and future prospects,"* Applied Microbiology and Biotechnology volume 104, 4281–4288, 2020.
- Gill Alexander, *"The Importance of Bacterial Culture to Food Microbiology in the Age of Genomics,"* Front Microbiol; 8: 777, 2017.
- Giese J., *"Color Measurement in Foods as a Quality Parameter,"* Food Technology, 54, pp. 62-64, 2000.
- Karajanagi Sandeep, Yoganathan Roshan, Mammucari Raffaella, Park Hyoungshin, Cox Julian, Zeitels Steven, Langer Robert, Foster Neil, *"Application of a Dense Gas Technique for Sterilizing Soft Biomaterials,"* Biotechnology and Bioengineering. 108, pp. 1716-25. 10.1002/bit.23105, 2011.
- Ribeiro Nilza, Soares Gonçalo, Santos-Rosales Víctor, Concheiro Angel, Alvarez-Lorenzo Carmen, García-González C.A., Oliveira A., *"A New era for sterilization based on supercritical CO_2 technology,"* Journal of Biomedical Materials Research Part B: Applied Biomaterials. 108. 10.1002/jbm.b.34398, 2019.
- Rimondini L., Fini M., Giardino R., *"The Microbial Infection of Biomaterials: A challenge for clinicians and researchers. A short review,"* Journal of Applied Biomaterials & Biomechanics; Vol. 3 no. 1: 1-10, 2005.
- Rózalska B, Sadowska B, Wieckowska M, Rudnicka W. Wykrywanie Biofilmu Bakteryjnego na Biomateriałach Medycznych, *"Detection of Bacterial Biofilm on Medical Biomaterials,"* Med Dosw Mikrobiol.,50(1-2):115-22. Polish. PMID: 9857621, 1998.
- Spilimbergo S, and Bertucco A, *"Non-thermal Bacterial Inactivation with dense CO_2,"* Biotechnol Bioeng. Dec 20;84(6):627-38. doi: 10.1002/bit.10783, 2003.
- STAATT, Technical Assistance Manual: State Regulatory Oversight of Medical Waste Treatment Technology. Report of the State and Territorial Association on Alternative Treatment Technologies, 2005.
- Suda I., Hajika M., Nishiba Y., Furuta S., Igita K., *"Simple and Rapid Method for the Selective Detection of Individual Lipoxygenase Isozymes in Soybean Seeds,"* J.Agri.Food.Chem, 43, 742-747, 1995.
- Von Woedtke Thomas, Kramer Axel, *"The Limits of Sterility",* GMS Krankenhaushygiene Interdisziplinär, Vol. 3(3), ISSN 1863-5245, 2008.
- Zhang Jian, Davis Thomas A., Matthews Michael A., Drews Michael J., *"Sterilization Using High-Pressure Carbon Dioxide,"* Journal of Supercritical Fluids The 38(3):354-372, 2006.
- Zwietering Marcel H., Jacxsens Liesbeth, Membré Jeanne-Marie, Nautae Maarten, Peterz Mats, *"Relevance of Microbial finished product testing in food safety management,"* Food Control, Volume 60, pp. 31-43, February 2016.

❐

4

Conventional Processes for Sterilization and Preservation

This chapter presents outlines of the processing technologies used for sterilization and preservation of various foods in the food industry, and for sterilization of biomaterials and medical devices in healthcare industries. Sterilization is a method used to get rid of all forms of microbial life including thermo-resistant spores in a material or an object, whereas preservation is the process that helps to kill only the vegetative form of bacteria but not the spores. Sterilization is different from disinfection or sanitization; the latter methods reduce rather than eliminate all forms of life and biological agents present.

4.1 Food Preservation

Food preservation can be defined as the process of treating and handling food in such a way as to stop or greatly slow down spoilage and prevent food-borne illnesses, while maintaining nutritional value, texture, and flavor. Food spoilage is directly related to food safety. A food preservation process renders a food product safe for consumption and enhances its shelf life. The shelf life of sterilized products is longer as compared to that of other preserved products. Food preservation and sterilization methods use moist heat using saturated steam under pressure, dry heat using hot air, membrane filtration, ionizing irradiation (e.g., gamma and electron-beam radiation), or by chemical (e.g., ethylene oxide) methods. The moderate heat treatment of pasteurization allows the destruction of pathogenic microorganisms present in their vegetative form and a large number of spoilage microorganisms.

M. Mukhopadhayay and A. Chatterjee, *Sterilization and Preservation*,
https://doi.org/10.1007/978-3-031-17370-7_4

There are many methods of preservation, some of which are centuries old, like pickling, shade drying, salting, fermenting, preserving in sugar concentrates, etc. However, the afore-mentioned methods and other conventional or emerging technologies alter one or all of the food characteristics, such as, the original taste, color, nutritional value, flavor, and aroma in some way or the other. This makes the processed product less consumer-friendly. In view of this, less severe preservation procedures are being sought in order to deliver products with enhanced shelf life and nutritional value. Most of the existing and emerging preservation techniques act by interfering with the homeostatic mechanisms that have been evolved by the microorganisms in order to survive extreme environmental stresses *[Gould, 1996]*. Preserved foods and preservation technologies directly impact the health of consumers; it is thus imperative to ensure high standards of sterilization and preservation of foods. The preservation techniques currently employed can be broadly classified based on the underlying principles of preservation, namely, (i) reduction in temperature, (ii) reduction in pH, (iii) reduction in water activity, and (iv) application of heat, as shown in *Figure 4.1.*

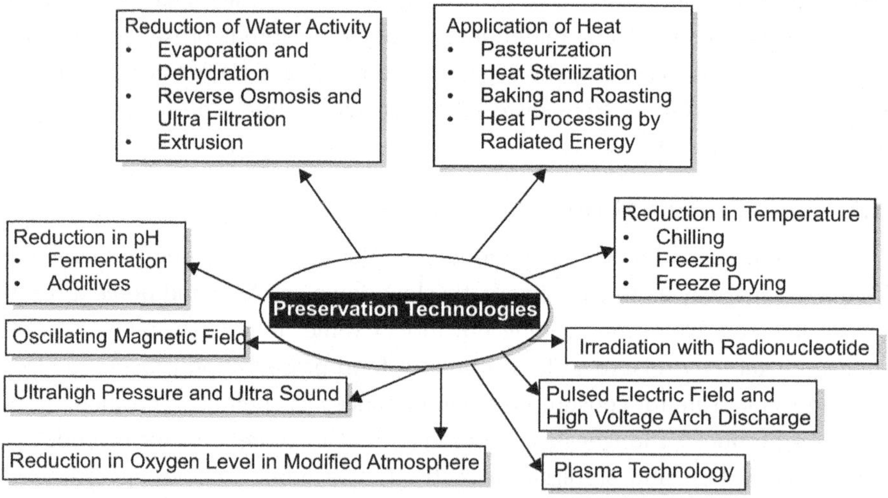

Fig. 4.1: Classification of preservation techniques with respect to underlying principles

4.2 Food Preservation Techniques

Preservation of food products entails the delay or prevention of microbial growth and consequent spoilage of foods. *Table 4.1* summarizes the key factors responsible for preservation by the existing and emerging technologies to achieve the desired goals and to enhance the shelf life.

Table 4.1: Key factors responsible for preservation techniques *[Gould, 1996]*

Goal	Key Factor for Preservation	Technique
Reduction or Inhibition of Growth	• Low Temperature • Low Water Activity • Restriction of Nutrient Availability • Lowered Oxygen • Raised Carbon Dioxide • Reduction in pH • Alcoholic Fermentation • Use of Preservatives	• Chill and Frozen Storage • Drying, Curing and Conserving • Compartmentalization in Water-in-Oil Emulsions • Vacuum and Nitrogen Packaging • Modified Atmosphere Packaging • Addition of Acids, Fermentation • Brewing, Vinification, Fortification • Addition of Preservatives, inorganic (sulphite, nitrite), organic, (propionate, sorbate, benzoate, parabens), antibiotic (nisin, natamycin)
Inactivation of Microorganisms	• Heating • Irradiating • Pressurizing • Electroporating • Manothermosonication • Cell Lysis • Disruption of Cellular Metabolic Activity • Structural Disruption	• Pasteurization and Sterilization • Ionizing Irradiation • Application of High Hydrostatic Pressure • High Voltage Electric Discharge • Heating and Ultrasonication at Slightly Raised Pressure • Addition of Bacteriolytic Enzymes (lysozyme) • Oscillating Magnetic Fields • $scCO_2$ Technology

4.2.1 Reduction in Temperature

Low temperature slows down the rate of all chemical reactions including those catalyzed by enzymes and it also decreases the fluidity of the cell membrane. Freezing the membrane prevents much of cellular metabolism by preventing enzymes in the membrane from functioning properly. Refrigerator temperature (~ 4°C) prevents the growth of many bacteria, extending the shelf life of many products. However, there are a large number of microorganisms still capable of slow growth at 4°C and these will eventually spoil foods. Refrigeration merely slows down the process of spoilage. Freezing a sample at or below-20°C stops all microbial growth. Low temperature, even freezing, is not damaging to most microorganisms and when brought up to suitable temperatures, the microbes will resume growth. In fact, microorganisms are well preserved at liquid nitrogen temperature (-196°C) and this is a common method of preserving bacterial strains in research laboratories. The following preservation techniques are based on application of this principle:

i) Chilling: Chilling is a preservation technique in which the temperature of a food is reduced to between -1°C and -8°C. It is used to reduce the rate of biochemical and microbiological changes, and hence to extend the shelf life of fresh and processed foods. It causes minimal changes to sensory and nutritional properties of food. The most significant effect of chilling on the sensory characteristics of processed foods is hardening due to solidification of fats and oils. Chemical, biochemical and physical changes during refrigerated storage may lead to loss of quality and may reduce shelf life *[Amit et al., 2017]*.

ii) Freezing: Freezing involves reduction of the temperature of a food below its freezing point and a proportion of water undergoes a change in state to form ice crystals. Long-term freezing requires a constant temperature of -18°C or less. However, freezing only slows the deterioration of food; it does not stop it, and while it may stop the growth of microorganisms, it does not necessarily kill them. Many enzyme reactions are only slowed by freezing. So it is often important to stop enzyme activity before freezing, either by blanching or by adding chemicals. The texture and taste of the original food is lost in this process *[Amit et al., 2017]*.

iii) Freeze Drying: In freeze drying the food particles are frozen rapidly to form ice crystals followed by removal of water to make the food dry. Freeze drying works by freezing the material and then reducing the surrounding pressure to allow the frozen water in the material to sublimate directly from the solid phase to gas. Freeze concentration of liquid food involves fractional crystallization of water to ice and subsequent removal of ice. Temperatures used are typically below –50°C. The drawback of the process is that texture and taste of the original food is lost *[Amit et al., 2017]*.

4.2.2 Reduction in pH

Weak organic acids are lipophilic. It is the lipid solubility of their undissociated forms that enables them to cross the microbial membrane and gain access to the cytoplasm of the cell. The pH value and the dissociation constant determine the proportion of acid that is in the undissociated form. At the pH value of most of the foods the microorganisms maintain an internal pH that is higher than that of their surroundings. On entering the cytoplasm, the undissociated acids tend to dissociate, delivering hydrogen ions along with the particular anion. The additional hydrogen ions may be exported by the cell from the surrounding to maintain a high internal pH. As this is energy demanding, the cell growth is restricted. If the energy demand is overcome the pH of the cytoplasm eventually falls to a level that is too low to sustain growth. Fermentation and additives are based on this principle.

i) **Fermentation:** Fermentation is a natural process through which microorganisms like yeast and bacteria convert carbohydrates, such as, starch and sugar into alcohol or acids. The alcohol or acids act as a natural preservative in fermented foods. Fermentation usually implies that the action of microorganisms is desirable. In food processing, fermentation typically means the conversion of carbohydrates to alcohols and carbon dioxide or organic acids using yeasts, bacteria, or a combination thereof, under anaerobic conditions. It may well be said that in fermentation processing the growth of the desirable bacteria inhibits the growth of the bacteria causing spoilage. Fermenting processes usually lower the pH of foods, and subsequently prevent growth of harmful microorganisms.

ii) **Additives:** Vinegar and citric acids have low pH, addition of these in food lowers the overall food pH which inhibits growth of microflora and are hence used as additives for preservation, such as, in preservation of meat, poultry, jam, jellies, pickles, etc. Bacterial growth is optimum around the neutral pH. Microorganisms, including yeasts, molds, and bacteria, are sensitive to a food pH. Very low or very high pH values will prevent microbial growth. However, very few foods have pH values low enough to completely inhibit the growth of microorganisms, especially yeasts and molds, which can tolerate lower pH conditions than most bacteria. Thus, an additional preservation technique is required, though food acidity or its pH value has a considerable influence on the selection of heat preservation processing conditions *[Amit et al., 2017]*.

4.2.3 Reduction in Oxygen Level in Modified Atmosphere

Controlled Atmosphere Storage (CAS) or Modified Atmosphere Storage (MAS) involves reduction in the concentration of oxygen and/or an increase in carbon dioxide concentration of the storage atmosphere surrounding a food to reduce the rate of respiration of fresh fruits and vegetables and also to inhibit microbial growth. CAS/MAS is to be used in combination with chilling and this enhances shelf life. The low levels of oxygen or high levels of carbon dioxide, which are needed to inhibit microorganisms, are harmful to many foods. An incorrect gas combination may change the biochemical activity of tissues, leading to development of off-odors and off-flavors.

The elimination of oxygen from the modified atmosphere pack alters the spoilage flora by preventing growth of strict aerobes and slowing the growth of the facultative anaerobes by restricting the amount of energy they can derive from the substrates. Many oxidative gram-negative bacteria like the *Pseudomonas* species are sensitive to carbon dioxide even in concentrations as low as 5%, while many lactic acid bacteria and yeasts are capable of growth even at 100% concentrations (even in CO_2 under pressure). Carbon dioxide – enriched food packets therefore result in a shift of spoilage flora from rapidly growing gram-negative one to a slower growing gram-positive association of strains *[Gould, 1996]*.

4.2.4 Reduction of Water Activity

All living entities require water. The removal of water (reduction of water activity to values between 0.60 and 0.86-0.90) from a sample inhibits the growth of microorganisms.

Water activity is indicative of the amount of water in the food that is available to microorganisms. Microflora have its specific minimum water activity value requirement, below which growth is no longer possible. The key application of water activity is the growth of bacteria, yeasts, and molds. In order to prolong the shelf life of food, without using refrigeration it is imperative to control either its pH levels or the level of water activity or a combination of both. Food with a water activity below 0.6 will not support the growth of osmophilic yeasts, which is a threat to high sugar products. *Clostridium botulinum*, the dreaded food poisoning bacterium, cannot grow at a water activity of 0.93 and below. However, *Staphylococcus aureus,* a common food poisoning organism, can grow in relatively low water activity level, such as, in cheese and fermented sausages stored above the recommended refrigeration temperatures.

Most food stuffs contain carbohydrates and proteins, and are therefore subject to nonenzymatic browning reactions. The Maillard's reaction is more pronounced at higher water activity values and reaches its maximum at water activity of 0.6 to 0.7. At low water activity values, foods that have high fat content acquire a rancid off-flavor after being stored for some time. The oxidation of fats and other food components shows a sharp decrease at water activity values below 0.2. Most enzymatic reactions are slowed down at water activity values below 0.8. Hence depending on the desired effect, the water activity needs to be controlled for chemical and enzymatic stability.

Water can be removed from foods by heating, evaporation, freeze drying or by addition of salt or sugar. The first three methods directly remove water from the sample and the last two rely on the hydrophilicity of salt and sugar to bind the water, limiting its availability to microbes. Before refrigeration, the salting of food was a common method of preserving it for prolonged periods.

Dried fruits and honey are other examples of low water-activity foods. There are microorganisms that have adapted to low water activity conditions and can grow under very dry conditions. Many of these are molds and yeasts, and they are the most common spoilage organisms found in jams, jellies, and other low water-activity foods. Pathogens, which have evolved to live on our bodies, cannot tolerate these conditions; so they are rarely a hazard in such foods.

i) **Evaporation and Dehydration:** Evaporation or concentration by boiling renders the partial removal of water from liquid foods by boiling off water in the form of vapor. It increases solid content of a food and hence preserves it by a reduction in water activity. The drawback of this process is that the aroma compounds that are more volatile than water are lost during evaporation, reducing the sensory characteristics of most concentrates. In fruit juices, this results in loss of flavor. Dehydration refers to the nearly complete removal of water from foods to a level of less than 5%. Dehydration involves simultaneous application of heat and removal of moisture from foods. Drying takes place within a matter of seconds at temperatures approximately 200°C. The main purpose of dehydration is to extend shelf life of foods by reducing water activity. However, the freshness, texture, color, flavor, and aroma of the original food cannot be retained *[Amit et al., 2017]*.

ii) **Reverse Osmosis and Ultra Filtration:** Reverse osmosis or ultra filtration selectively removes water through a semi-permeable membrane, but cannot act as a preservation process unless combined with some other preservation technique. Moreover, in these processes the microbial spores remain in a dormant form and are not inactivated. As a result, on return of favorable conditions they resume growth. Nevertheless, the process of concentration may act as a good preservation process in some cases *[Mohammad et al., 2012]*.

iii) **Extrusion:** Extrusion is a process that combines several unit operations including mixing, cooking, kneading, shearing, shaping and forming. The raw materials are forced to flow under controlled conditions along the length of the extruder barrel and through a die assembly at a fixed throughput. During extrusion the product is cooked and mixed by three separate energy sources: mechanical energy (shear caused by the screw elements), thermal energy that comes from the heating system, and self heating due to the melt viscosity in the barrel. The main method of preservation of hot and cold extruded foods is by reduction of water activity. It is the most common technology used for all snack foods and 'ready to eat' fried foods. However, it causes denaturation of proteins, melting of fats, hydration of starches, plastification, and expansion of food structure *[Shelar and Gaikwad, 2019]*.

4.2.5 Application of Heat

High temperature kills microbes by causing lysis of the membrane or denaturation of critical enzymes. Application of heat is a preservation technique in which foods are heated at a sufficiently high temperature and for a sufficiently long time (121°C for 15-30 minutes) to destroy microbial and enzyme activity. As a result, sterilized foods have a shelf life in excess of six months at ambient temperatures. Application of heat can be generally rendered using moist/wet heat or dry heat needed for food preservation. Dry heat involves incubation in an oven-like environment, while moist heat utilizes steam under pressure, and the latter is more effective. In case of moist/wet heat, temperatures used are in the range of 121–129°C along with application of pressure. For dry heat, temperatures used are in the range of 176–232°C. However, the duration of treatment is longer in comparison to wet heat. The use of dry heat requires specific time, temperature, and other process parameters to be determined individually for each type of material that are being sterilized. However, for heat resistant materials higher temperatures and shorter time of treatment may be used.

In case of wet heat, the heat transfer to the material is much better; as a result the overall exposure time is shorter and lower temperatures can be used. Dry heat on the other hand is not suitable for plastics owing to low thermal transmission properties of plastics, thus making it an undesirable process for plastic packaged food materials *[Chakraborty, 2006]*.

Water has a very high heat capacity and moist air is capable of holding more heat than dry air. The moist heat technique is therefore more effective; because it increases the rate of heat penetration into a substance. With the dry heat technique a higher temperature or longer time of exposure is necessary to obtain the same amount of killing as that of the moist heat. Severe heat treatment causes severe damage to the nutritional qualities and in some cases may even impart a cooked flavor to the food product.

The goal of heat treatment is to bring the target population down to some acceptable level. In addition, each species of microbe also has its own characteristic resistance to heat, with some bacteria being much more heat tolerant than others. A final factor influencing the effectiveness of a heat treatment process is the composition of the environment, surrounding the microbe. High salt and acidic environments increase the rate of killing at a given temperature due to the damaging effects that the salt and acid have on the cell. Conversely, fats and proteins in a solution have a protective effect.

i) **Pasteurization:** Pasteurization refers to a moderate heat treatment, leading to inactivation of microorganisms without significant degradation of food products. The difference between sterilization and pasteurization is that the latter does not kill spores. Pasteurization is the first and still most common method for preservation by application of heat. It was originally developed to prevent the spoilage of wine; it is commonly used for milk, and eliminates the transmission of *Coxiella burnetti, Mycobacterium tuberculosis, Brucella, Staphylococcus, Salmonella,* and *E. coli* strain O157:H7. In the original batch pasteurization method, the food was heated at 66°C for 30 minutes. However, the most modern application uses flash pasteurization, which is a treatment at 72°C for 15 seconds. The shorter incubation time allows easier automation and less damage to food during processing. Also, a higher temperature is more effective at killing *Coxiella burnetti.*

Pasteurization, in general, is used for liquid foods, though typically associated with milk. There are two widely used methods to preserve milk: high temperature short time (HTST) and ultra-high temperature (UHT). HTST is by far the most common method. Milk simply labeled "pasteurized" is usually treated with the HTST method, whereas milk labeled "ultra-pasteurized" must be treated with the UHT method. It may be noted that the UHT- processed milk is sterile but not pasteurized, and has longer shelf life. HTST involves holding the milk at a temperature of 75°C for 15-20 seconds. Whereas, UHT involves holding the milk at a temperature of higher than 135°C for at least two seconds [*Gedam et al., 2007*].

Pasteurization not only eliminates pathogens, but also greatly decreases the number of spoilage organisms. Pasteurization is used extensively in treating many other food products including beer, wine, yogurt, juices, and cheese. In low acid foods (pH> 4.5, for example milk) it is used to minimize possible health hazards from pathogenic microorganisms and to extend the shelf life of foods for several days. In acidic foods (pH< 4.5, for example bottled fruit) it is used to extend the shelf life by delaying spoilage by microorganisms and/or enzyme inactivation. In pasteurized foods there are only minor changes to the nutritional and sensory characteristics of most foods. However, the shelf life of pasteurized foods is usually extended by a few days or weeks *[Cappozzo et al., 2015].*

ii) **Heat Sterilization:** Heat sterilization is the process of heating to a high enough temperature (usually more than 100°C) for specific time to kill almost all bacteria, for example, milk or fruit juice sterilized at more than 100°C can be stored at room temperature for a long period of time.

However, for avoiding milk from boiling at the sterilization temperature, a high operating pressure is needed to be employed. In general, heat sterilization can be carried out by two methods:

A. In the conventional method, the heat treatment is usually carried out at 105-110°C for 30-45 min after packaging is done and so it is also known as 'canning' or 'in-bottle sterilization'. There are two problems associated with the conventional method of heat treatment, namely: (i) heat has to be passed through the container first before it goes into the liquid food, e.g., milk or fruit juice and (ii) a ring of precipitated solid is likely to get deposited on wall of the bottle at edges of the liquid surface due to excessive foam formed during bottle filling operation. To avoid this, it is necessary to rotate or shake the bottle vigorously during the heating and cooling cycle, and the method can be designed to operate as a batch or continuous process.

B. In the other method, known as the Ultra High Temperature (UHT) or aseptic method, the packaging is done after heat treatment. The latter is also known as the Ultra High Temperature Short Time (UHTST) or Very High Temperature Short Time (VHTST) process in which the processing is carried out at still higher temperature at 135-150°C for 4-15 seconds at a pressure of 4-6 bar prior to packaging the sterilized product aseptically [*Gedam et al., 2007*]. It may be noted that the length of holding time is much less than the HTST pasteurization method. The UHT sterilization process causes much less chemical change than the conventional sterilization process. The heat treatment can be carried out as a continuous process with indirect heat of steam in a heat exchanger or by direct steam injection for heat treatment. The direct steam injection can increase the temperature of the liquid food within fraction of a second and the holding time required is very less. However, it is necessary to maintain a sufficient back-pressure, say minimum 2 bar on the product, not only to prevent boiling at the highest temperature, but also to prevent the separation of dissolved gases which are normally present in the product. This ultimately helps to minimize fouling *[Dash, 2020].*

iii) **Baking and Roasting:** Baking and Roasting are essentially the same unit operation, as both use heated air to remove moisture and alter the eating quality of foods. The terminology baking is applied to flour-based foods or fruits, and roasting of meats, nuts, and vegetables. This method cannot be used for preservation of fresh fruits and vegetables. Heat-labile proteins and pigments get denatured in this method.

iv) **Heat processing by Radiated Energy:** Heat Processing can be done by direct or radiated energy, e.g., Dielectric, Ohmic, and Infrared. Dielectric energy and infrared energy are two forms of electromagnetic energy and are transmitted as waves which penetrate the food, then absorbed, and are converted to heat. Ohmic heating uses electrical resistance of foods to directly convert electricity to heat. This type of processing is not uniform and also shelf life enhancement is not significant.

4.2.6 Irradiation with Radionucleotide

It is known that γ-rays break the chemical bonds when absorbed by materials giving the products of ionization, which may be electrically charged or neutral. These then further react to cause changes in an irradiated material known as radiolysis. It is these reactions that cause the destruction of microorganisms, insects, and parasites during food irradiation. The limitation of this process is that it is not permitted in all countries. Moreover, an isotope radiation plant is required, the permission for which is hard to acquire. Also, the toxins as a result of inactivation of microbes and insects remain in the food.

Radionucleotide sources are used to generate Pulsed X-Rays, which use a solid state-opening switch to generate electron beam X-ray pulses of high intensity. Neutron bombardment of Co-59 and Cs-136 as a fission fragment of a nuclear power reactor operation produces radionuclides Co-60 and Cs-137. These emit γ-radiation of discrete energy. Alternatively, electrically driven radiation sources that can be switched off when no longer needed are easier to handle. Linear Induction Electron Acceleration (LIEA) generates energetic electrons and X-rays that deliver ionizing dose to food products. These radiations are provided by an electron accelerator system. The electron source used for this purpose is almost always a thermionic gun, the cathode of which is raised to a sufficiently high temperature. LIEA can deliver dose rates that are higher in magnitude as compared to Co-60 sources *[Miller, 2005]*.

4.2.7 Pulsed Electric Field and High Voltage Arch Discharge

Pulsed electric field makes use of electric discharge (12-35kV cm^{-1}) in short pulses (1-100μs) for destruction of microorganisms and enzymes. The vegetative cells are killed in this process, and the sensory and nutrient characteristics are preserved. The drawback of this treatment is that it is effective only in combination with heat and it is unable to destroy spores and enzymes. The process is, however, difficult to use with conductive materials.

The High Voltage Arch Discharge (HVAD) method involves the application of electricity to liquid food items in order to pasteurize them by generating intense waves and electrolysis, by means of rapid discharge through an electrode gap, thereby inactivating the microorganisms. This process depends not only on the voltage applied, but also on the type of microorganism, initial concentration of cells, volume of the media used, distribution of chemical radicals, and electrode material *[Jambrak et al., 2017]*.

4.2.8 Ultrahigh Pressure and Ultrasound

Ultrahigh Pressure (UHP) offers interesting possibilities for food processing ranging from extraction of plant compounds, restructuring of foods, and rapid formation of small ice crystals. High Pressure treatments make use of pressures as high as 658MPa for 10 minutes (typically pressures in the range of 350-1000MPa are employed). This can destroy microbes since the microbes in log phase are baro-sensitive, and the sensory characteristics are also conserved. But this treatment alters protein structures and the enzymes largely remain unaffected. Pressures between 300 and 600MPa can inactivate yeasts, molds, and most vegetative bacteria including most infectious food-borne pathogens. Thus, pressure is a potential alternative to heat pasteurization as pressure leaves small molecules, such as, many flavor compounds and vitamins intact. Bacterial spores can often be stimulated to germinate by pressures of 50-300MPa. Germinated spores can then be killed by relatively mild heat treatments or mild pressure treatments. Exponentially growing cells are more sensitive to pressure than cells in the stationary phase. Stress might be induced during the stationary phase (e.g., through starvation or acidification). A perturbation of the bacterial membrane is almost always involved during pressure treatment *[Smelt, 1998]*. The only drawback of this process is the very high pressures involved.

Ultrasound (10 – 1000 W cm^{-2}, frequency up to 2.5 MHz) produces very rapid localized changes in pressure and temperature that cause shear disruption, thinning of cell membranes, localized heating, and free radicals, which have lethal effect on microorganisms. The use of this treatment could cause unwanted modification in structure and texture; the production of free radicals can cause rancidity.

4.2.9 Plasma

Plasma sterilization operates based on its specific active agents, which are ultraviolet (UV) photons and radicals (atoms or assembly of atoms with unpaired electrons, therefore chemically reactive, e.g., O^- and OH^-, respectively). An advantage of the plasma method is the possibility, under appropriate conditions, of achieving sterilization at relatively low temperatures (≤50°C). However, uniformity of the process cannot be guaranteed and the process is time-dependent. The longer the processing time, the better is the sterilization achieved *[Moisan et al., 2002]*.

The antimicrobial effects of light at UV wavelengths are due to absorption of the energy by highly conjugated double carbon bonds in proteins and nucleic acids, which disrupts cellular metabolism; the range of the light is from UV wavelengths of 200nm to infrared wavelengths of 1000nm. This treatment is however ineffective against spores and its effect is limited only to the surface of the food products.

4.2.10 Oscillating Magnetic Field

Oscillating magnetic field (OMF) is applied for preservation and sterilization of food in which pulses are applied to the food in the form of decaying or constant amplitude sinusoidal waves. The magnetic fields applied may be homogeneous (due to uniform magnetic field intensity) or heterogeneous (as the magnetic field intensity is inversely proportional to distance from the coil) depending on the food type *[USA-FDA, 2011]*. Preservation of foods with OMF is performed by sealing the food in a plastic bag and subjecting it to 1 to 100 pulses in an OMF with frequency between 5 and 500 kHz, and temperatures at 0 to 50°C, with exposure time ranging from 25 to 100 milli-seconds.

The restriction or arresting of microbial activity in most of the traditional and novel food preservation techniques is based on a combination of several preservation factors called the hurdles. The microorganisms present in food are unable to overcome these hurdles. As a result, the food is rendered safe for consumption and has prolonged shelf life. In most cases multi hurdle is used to avoid harshness that would have otherwise been used if only one hurdle treatment is used. The use of $scCO_2$ in the area of food sterilization and preservation is one of the potent emerging technologies and will be discussed in detail in Chapters 6-8. *Figure 4.2* shows various emerging novel technologies for food sterilization and preservation.

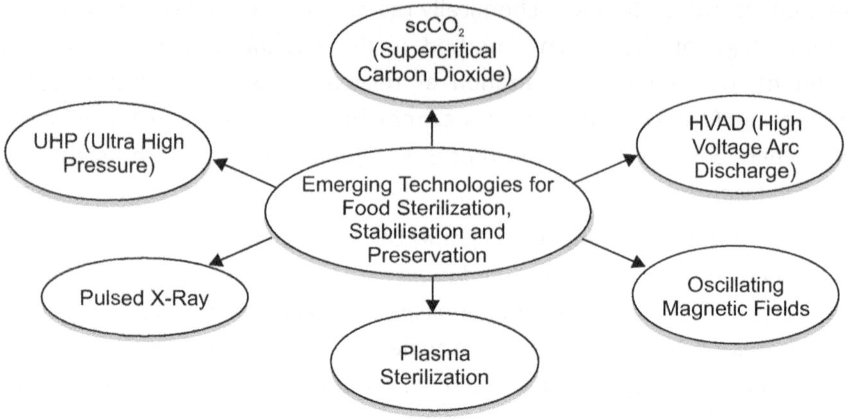

Fig. 4.2: Emerging technologies for food sterilization and preservation

4.3 Sterilization of Biomaterials and Medical devices

As healthcare technology has evolved over the years, it has immensely benefitted the humanity by facilitating longer life, better health, and faster cure. Biomaterials which find applications either in medical devices or as implants, come into human contact and must first be sterilized and maintained in sterile condition within the standard limit until usage, to ensure success. Sterility of materials is usually assured by less than one in one million devices having a microbial presence (i.e., SAL of 10^{-6}). Sterilization is the process of making the materials free from bacteria or other living microorganisms.

Most biomaterials and medical devices used in healthcare facilities are generally made of materials that are heat and moisture stable. However over the recent years, there has been an increase in the use of heat sensitive polymeric biomaterials for medical devices and instruments. Sterilization of biomaterials has been conventionally achieved using thermal, chemical or radiation methods. If these materials are heat resistant, the recommended sterilization process is thermal sterilization, because it has the largest margin of safety due to its reliability, consistency, and lethality. For disinfection of glass and metallic devices, thermal sterilization treatments are most commonly used. However, reprocessing of heat- and moisture-sensitive items requires use of a low-temperature sterilization technology. In recent years several sterilization technologies have been developed, using chemicals, e.g., ethylene oxide (ETO) gas, hydrogen peroxide (H_2O_2) gas plasma, peracetic acid (PAA) immersion, and ozone, among others *[Soares et al., 2019]*. These chemicals as sterilant are used to sterilize heat-sensitive medical devices at relatively low temperatures. Ethylene oxide (ETO) gas has been used for heat- and moisture-sensitive medical devices. The use of gamma irradiation and electron beam (E-Beam) treatment offers advantages for scale-up, and are typically preferred as methods for industrial level sterilization. While patient safety is of primary consideration in the choice of the appropriate method to ensure sterility, emphasis is given to the impact of the harshness of the sterilization method used on the biomaterial and any potential changes in its functional performance. This section presents sterilization technologies used in healthcare facilities and highlights their applicability in the processing of biomaterials and medical devices.

4.3.1 Thermal Sterilization

Heat in general can improve and enhance the microbial effectiveness of any method of sterilization. It may adopt two kinds of heating, namely, wet and dry. Steam and dry heat sterilization function similarly in that both have the ability to sterilize virtually all organisms with no toxic residues or waste. However they have differences, namely, steam sterilization may deform, corrode or wet the materials, whereas dry heat can degrade and melt many heat-sensitive materials and devices, though dry heat has excellent penetration capabilities compared to steam heating.

Steam sterilization involves exposure of a material to direct steam in an autoclave at the required temperature and pressure for a specified time. It generally uses 121°C for 15 to 30 min to kill resistant bacteria. An autoclave is a large steel vessel or chamber that circulates steam at high temperature and pressure to sterilize various items. Steam sterilization is one of the most economical methods of sterilization, while still having a short cycle time. On the other hand, dry heat sterilization uses conductive heating, with little or no water vapor and requires higher temperatures to kill resistant spores. The most common time-temperature relationships for sterilization with hot air sterilizers are 170°C for 60 minutes, 160°C for 120 minutes, and 150°C for 150 minutes. Over the years, both of these thermal methods have been practiced in hospitals and medical facilities to sterilize mostly metallic medical instruments. However, these methods are not suitable for thermally sensitive biomaterials *[White et al., 2006]*.

4.3.2 Ethylene Oxide (ETO) Gas Sterilization

It is normally used for heat- and moisture–sensitive biomaterials, mostly for packaged medical devices, or materials inside enclosed chambers. ETO is a colorless gas that is flammable, explosive, and carcinogenic. It is able to penetrate the material to achieve the desired sterilization and may be combined with other gases like CO_2 or hydro chloro fluoro carbons (HCFC), requiring a pressure vessel. The effectiveness of ETO sterilization depends on four essential parameters, namely: gas concentration (which ranges from 450 to 1200 mg/l), temperature (between 37°C and 63°C), relative humidity (in the range of 40 to 80%) as water molecules carry ETO to reactive sites, and exposure time from 1 to 6 hours depending on the operating parameters. The microbicidal activity of ETO is considered to be the result of alkylation of protein, DNA, and RNA. Alkylation or the replacement of a hydrogen atom with an alkyl group, within cells prevents normal cellular metabolism and replication. After the sterilization process, it requires thorough aeration to eliminate absorbed ETO, as it can leave toxic residues on products *[Mendes et al., 2012]*. The main drawbacks are the lengthy cycle time, the cost, and its potential hazards.

4.3.3 Vaporized Hydrogen Peroxide (VHP) Sterilization.

It is also known as hydrogen peroxide gas sterilization in which hydrogen peroxide (H_2O_2) kills microbes by oxidizing amino acids and proteins. Vaporized hydrogen peroxide (VHP) sterilization is a low-temperature vapor process that has traditionally been used for sterilization of reusable heat-sensitive medical devices in patient care facilities. It involves three steps, namely: (i) liquid H_2O_2 gets converted into vapor, (ii) the vapor fills the sterilization chamber for contacting all surfaces and penetrating lumens, and (iii) after sterilization, the vapor is removed from the chamber with the help of vacuum and converted into water and oxygen.

The VHP generator may use deep vacuum to pull liquid hydrogen peroxide solution in water (30-35% concentration) from a disposable cartridge for vaporization into a heated vaporizer and then into the sterilization chamber for contacting the device surfaces to be sterilized with VHP. Alternatively, VHP may be generated by passing a carrier gas, such as, air using either a slight negative pressure or slight positive pressure to circulate the vapor in the sterilizing chamber at a predetermined concentration in the air, typically from 140 ppm to 1400 ppm, depending on the infectious agent to be cleared. These vapors effectively remove micro-organisms that may be present, sterilizing the enclosure. A VHP sterilization cycle typically requires less time than alternative forms of sterilization, such as, ETO sterilization. Once the sterilization cycle is completed, the vapor is removed from the chamber. The generator then reverses the process, breaking down the hydrogen peroxide vapor into environment-friendly elements. This method has one of the lowest cycle times, resulting in the ability to sterilize equipment in high volume batches. No ventilation is necessary for the vaporized hydrogen peroxide sterilization process. VHP offers several appealing features that include rapid cycle time (e.g., 30-45 minutes), low temperature, environmentally safe by-products (H_2O, oxygen), good material compatibility, and ease of operation, installation and monitoring. VHP has limitations including that cellulose cannot be processed, nylon becomes brittle, and VHP penetration capabilities are less than those of ETO *[Pottage et al., 2012]*.

4.3.4 Hydrogen Peroxide Gas Plasma Sterilization

It is a low temperature sterilization process, commonly used to sterilize heat-sensitive devices. It utilizes low temperature hydrogen peroxide gas plasma within a chamber to kill all living microorganisms on medical and dental equipment, including bacteria, spores, viruses and fungi. Gas plasmas are generated in an enclosed chamber under deep vacuum using radio frequency or microwave energy to excite the gas molecules and produce charged particles, many of which are in the form of free radicals. A free radical is an atom with an unpaired electron and is a highly reactive species. The free radicals (e.g., hydroxyl and hydroperoxyl) generated in the gas plasma are microbicidal. If any moisture is present on the objects, the vacuum will not be achieved and the cycle is rendered ineffective.

In this process, liquid H_2O_2 is heated up in a vaporizer in order to turn it into gas and fed into the sterilizer. The H_2O_2 vapor diffuses through the chamber (in 50 minutes), exposes all surfaces of the load to the sterilant, and initiates the inactivation of microorganisms. Once that has been accomplished, the hydrogen peroxide gas is heated to an even higher temperature under vacuum, at which point it turns into plasma. Once the sterilization cycle is completed, the vapor is removed from the chamber by vacuum. In the final stage of the process the sterilization chamber is returned to atmospheric pressure by introduction of high-efficiency filtered air and any residual H_2O_2 is converted to water and oxygen. The by-products of the cycle (e.g., water vapor, oxygen) are nontoxic and eliminate the need for aeration, making operation safe for both medical staff and the environment. Thus, the sterilized materials can be handled safely, either for immediate use or storage. The process operates in the range of 37-44°C and has a cycle time of 75 minutes. Although plasma sterilization is a more expensive method, it is highly effective and ideal for moisture-and heat-sensitive medical tools *[Boiano and Steege, 2015]*.

4.3.5 Peracetic Acid Sterilization

This method relies on the biocidal effect of this acid to remove the surface contaminants on medical instruments. Peracetic acid (PAA) is a highly biocidal oxidizer that maintains its efficacy in the presence of organic soil. It is used to chemically sterilize medical, surgical, and dental instruments (e.g., endoscopes, arthroscopes). The sterilant is 35% PAA, an anticorrosive agent and is supplied in a single-dose container. The concentrated PAA is diluted to 0.2% with filtered (0.2 mm) water at a temperature of approximately

50°C. The diluted PAA is circulated within the chamber of the sterilizing machine and pumped through the channels of the endoscope for 12 minutes, decontaminating exterior surfaces, lumens, and accessories *[Yoganarasimha et al., 2014]*.

4.3.6 Ozone Immersion Sterilization

It is one of the recent methods used to sterilize reusable medical devices. Ozone is a sterilant as it consists of O_2 with a loosely bonded third oxygen atom that is readily available to oxidize other molecules. This additional oxygen atom makes ozone a powerful oxidant that kills microorganisms. The sterilant is converted back to oxygen and water vapor at the end of the cycle by passing it through a catalyst before being released into the environment. The duration of the sterilization cycle is about 4 h and 15 m, and it occurs at 30-35°C. Microbial efficacy has been demonstrated by achieving a SAL of 10^{-6} for a variety of microorganisms. Though this method operates at low temperature, but it has its own limitations, like it is found to alter morphology, structure, and surface properties of different organic polymers *[Premnath et al., 1996]*. The ozone process is compatible to several common materials including stainless steel, titanium, anodized aluminum, ceramic, glass, silica, PVC, Teflon, silicone, polypropylene, polyethylene, and acrylic. However, ozone is known to be an air pollutant which is very harmful to breathe and also it damages the crops in field and other vegetation.

4.3.7 High Energy-Gamma Irradiation

It is commonly used for sterilization of medical devices, microbial reduction of foods, cosmetics and their packaging, and the disinfestation of agricultural products, usually using high-energy photons emitted from an isotope source (e.g., Cobalt 60) as the radionuclide. These photons can create ionization (electron disruptions) in any material that it encounters. Sterilization by gamma radiation occurs by the release of high-energy photons that interact with substances. The common dosage used for medical plastics is 15–25 kGy, with 25 kGy being the standard dose applied. In living cells, these disruptions result in damage to DNA and other cellular structures. These photon-induced changes at the molecular level may cause the death of the organism or render the organism incapable of reproduction. This effect is useful in killing bacteria, insects, or other living contaminants which may exist in or on a product. In conjunction with its effect on DNA, the free radicals also have the ability to crosslink with polymers, which can deteriorate the properties of the devices and equipment, therefore placing a limitation on the use of gamma radiation as a means of sterilization. This is often detected by discoloration and hardening of plastics and a reduction in the mechanical strength of the materials *[IAEA, 2008]*.

4.3.8 Electron Beam (E-Beam) Sterilization:

It uses high power accelerators to provide a higher dosing of ionization to create a high-power electron beam compared to gamma irradiation sterilization. The process by which electron beam sterilization works is very similar to that of gamma radiation, where high-energy electrons are absorbed onto the material and produce secondary electrons. Similarly, the secondary electrons are those that eliminate the pathogens and can also alter material characteristics. The process of electron beam sterilization is faster (seconds to minutes) than gamma sterilization (minutes to hours) because of a very high delivery rate of dosage. There is no radioactivity involved. In electron beam sterilization, boxes of medical devices are put on a conveyor in a single layer. As they pass through the beam, electrons penetrate the cardboard box and all the medical devices in their individual packages inside the carton. Harmful microorganisms are completely inactivated with minimal effect on the medical devices. As the electrons penetrate the products, the radiation dose diminishes, less radiation leavs the box than entered. E-Beam provides the most cost-effective radiation sterilization for low to medium density medical devices *[Silindir and Özer, 2009]*.

4.3.9 Microwave Sterilization

Microwaves are used for sterilization of compatible medical devices (which do not melt), e.g., soft contact lenses, dental instruments, dentures, and urinary catheters, etc. Microwave sterilization is due to radio-frequency waves, which are usually generated at a frequency of 2450 MHz. The intermolecular friction derived from the vibrations of microwaves generates heat or it may cause a non-thermal lethal effect and so the microwaves are an effective microbicide. It is found that higher power microwaves in the presence of water may be needed for sterilization within a very short time.

4.4 Safety and Effectiveness of the Processes

In spite of low operating temperatures of the recently developed technologies as outlined above, they have their other limitations. Either they have been found to be corrosive, carcinogenic, and hazardous or they may distort morphology, structure, and surface properties of different organic biomaterials and polymeric devices *[Premnath et al., 1996; Lerouge et al., 2001; Silindir and Özer, 2009]*. *Table 4.2* summarizes the main advantages and disadvantages of the above mentioned sterilization techniques *[Rutala and Weber, 2008; Qiu et al., 2011]*.

Radiation-based method of sterilization occurs at dosages ranging from 8 to 35 kGy. Hence, it poses two main safety concerns, firstly there is possible lethal exposure to radiation and ozone. The lethal dosage of radiation to humans is about 0.01 kGy with an exposure time of less than a second in some cases. Thus safety measures are needed in radiation-based sterilization processes to prevent lethal exposure to workers and patients, such as, provision for considerable amounts of shielding and robust interlocks. In addition to exposure to radiation, the production of ozone is another issue for concern, as ozone is a toxic gas that is formed when radiation comes into contact with oxygen, causing first the formation of free oxygen radicals and successively ozone. Appropriate safety measures to prevent ozone inhalation include ozone monitors and adequate air ventilation. Sterilization by chemical methods is more involved and specific to the product and chemical used than methods, such as, irradiation. Environmental protection concerns related to ETO –based processes lead to regulatory control demanding for adoption of alternative, safer sterilization techniques. Oxidizing methods of sterilization using either hydrogen peroxide or ozone provide less of a threat with respect to human exposure; however, several safety precautions must still be implemented while following them. Hydrogen peroxide process begins with a source that is highly concentrated and extremely hazardous. Ozone, on the other hand, must be produced *in situ,* since it is metastable and cannot be stored. For harnessing the biocidal properties of either compound, the equipment and devices must be chosen on the basis of resistance to oxidation and corrosion, and mechanical strength, as often deep vacuum is required.

The sterilization process using a sterilizing agent must ensure that it will not affect the products safety, and also it should address the following aspects for its effectiveness and performance:

i) The effects of the process variables on the product and packaging materials, and the specified requirements for the sterility of the target material.

ii) The effectiveness of the sterilization process based on the inactivation of microorganisms to enable the process to perform in the most difficult-to-sterilize condition of the material.

iii) The tests of sterility for the biological safety of the product following exposure to the sterilizing agent.

iv) A post-treatment as part of the sterilization to reduce the level of residues to meet specific limits.

v) Identification of key process parameters and the routine monitoring and control of the process even after validating the sterilization process.

Table 4.2: Advantages and limitations of current techniques of sterilization of biomaterials, *[adapted from Perrut, 2012]*

Technology	Advantages	Disadvantages
Dry heat/steam	Nontoxic, Inexpensive, rapidly microbiocidal, short treatment time, good penetration	Deletorious for heat and/or moisture-sensitive materials
Ethylene oxide	Compatible with many biomaterials, low temperature setting for heat and/or moisture-sensitive materials.	Not suitable for cellulose (paper), linens and liquids, not suitable for devices with lumens
Peracetic acid	No activation required, odour or irritation not significant	Materials compatibility concerns, limited clinical use (only for immersible instruments/materials), point of use system, no long term sterile storage
Gamma Irradiation	Nontoxic, good panetration	Deleterious to polymers and biological materials
E-Beam	Nontoxic, short treatment time	Deleterious to polymers and biological materials, limited penetration distance

For past decades very high pressure treatment has been used for sterilization as an alternative to dry heat or chemicals treatment which generally degrades the product quality, especially in Japan where irradiation was never accepted. However its high operating cost is incompatible with most markets in view of the requirement of very high hydrostatic pressure (400-800MPa) and prolonged treatment times that are needed for efficient sterilization. In recent years, much attention has been paid to the $scCO_2$ treatment for inactivation of spores and virus that are known to be highly resistant to sterilization processes involving heat, radiation, and chemical agents. Presence of a strong oxidant additive to CO_2 such as, hydrogen peroxide, tert-tributyl hydroperoxide (t-TBHP) or mixture of both, or peracetic acid or trifluoroacetic acid even at very low concentration, results high sterilization efficiency even at mild temperature. A combination of a pulse electric field pre-treatment followed by $scCO_2$ treatment at 40°C is found very efficient on spores although each of these processes has no significant effect when operated alone *[Perrut, 2012]*.

References

- Amit Sadat Kamal, Uddin Md. Mezbah, Rahman Rizwanur, Islam S. M. Rezwanul, Khan Mohidus Samad, *"A review on Mechanisms and Commercial aspects of Food Preservation and Processing,"* Agriculture & Food Security, 6:51, 2017.
- Boiano JM, Steege AL, *"Ethylene Oxide and Hydrogen Peroxide Gas Plasma Sterilization: Precautionary Practices in U.S. Hospitals,"* Zentral Sterilization (Wiesb); 23(4):262-268, 2015.
- Cappozzo Jack C., Koutchma Tatiana, Barnes Gail, *"Chemical characterization of milk after treatment with thermal (HTST and UHT) and non-thermal (turbulent flow ultraviolet) processing technologies,"* J. Dairy Science, Vol. 98, Issue 8, pp. 5068-5079, 2015.
- Chakraborty Anuradha, *"Sterilization and Stabilization of Food Products Using Supercritical Carbon Dioxide,"* Ph.D. Dissertation, Indian Institute of Technology, Bombay, 2006.
- Dash S.K, *Course on Dairy and Food Engineering,* College of Agricultural Engineering and Technology, 2020.
- Dillow A.K., Dehghani F., Hrkach J.S., Foster N.R., Langer R., *"Bacterial inactivation by using near and supercritical carbon dioxide,"* Proc. National Academic Science, U.S.A. 96 (18) pp.10344–10348, 1999.
- Gedam, K, Prasad R, Vijay V.K., *"The study on UHT Processing of Milk : A versatile option Rural Sector"* World Journal of Dairy & Food Science 2 (2), pp 49-53, 2007.
- Gould GW., *"Methods of Preservation and Extension of Shelf life,"* Int. J. Food. Microbiology, 33, pp.55-64, 1996.
- Harrington Roger E., Guda Teja, Lambert Bryron, Martin Jeffrey, *"Sterilization and Disinfection of Biomaterials for Medical Devices"* Section 3.1.4, The Medical Product Life Cycle,1432 Chapter 3.1.4 Sterilization and Disinfection of Biomaterials for Medical Devices, 2020.
- International Atomic Energy Agency, Vienna, *"Trends in radiation sterilization of health care products,"* pp.5-25, 2008.
- Jambrak Anet Rezˇek, Vukusiˇ c´ Tomislava, Donsi Francesco, Paniwnyk Larysa, Djekic Ilija, *"Non-thermal processing in food applications: A review,"* International Journal of Food Science and Nutrition, Vol 2; Issue 6; pp.171-180, 2017.
- Lerouge S., Wertheimer M.R., Yahia L., *"Plasma sterilization: a review of parameters, mechanisms, and limitations,"* Plasmas Polym. 6 (3) pp.175–188, 2001.
- Mendes G.C., Brandao T.R.S., Silva C.L.M., *Ethylene oxide (EO) sterilization of healthcare products,* in: S. Lerouge, A. Simmons (Eds.), *Sterilization of Biomaterials and Medical Devices,* Woodhead Publishing Limited, Cambridge, UK, pp. 71–96, 2012.

- Miller R.B, *"Overview of Electron Accelerator Systems, In: Electronic Irradiation of Foods,"* Food Engineering Series. Springer, Boston, MA. 2005, https://doi.org/10.1007/0-387-28386-2_5.
- Mohammad Abdul, Ng Ching, Ying Pei Lim, Ng Gen., *"Ultrafiltration in Food Processing Industry: Review on Application, Membrane Fouling, and Fouling Control,"* Food and Bioprocess Technology. 5, 2012. 10.1007/s11947-012-0806-9.
- Moisan Michel, Barbeau Jean, Crevier Marie-Charlotte, Pelletier Jacques, Philip Nicolas, Saoudi Bachir, *"Plasma Sterilization Methods and Mechanisms,"* Pure Appl. Chem., Vol. 74, No. 3, pp. 349–358, 2002.
- Online, *"Sterilization methods overview gamma, e-beam, eto, steam & dry heat validations, Life Science Outsourcing,"* Inc.Available: http://www.lso-inc.com/sterilizationvalidation-services/sterilization-methods-overview.html, Accessed date: 30 May 2017.
- Perrut M., *"Sterilization and virus inactivation by supercritical fluids (a review),"* J. of Supercritical Fluids, 66, pp. 359-371, 2012.
- Premnath V., Harris W.H., Jasty M., Merrill E.W., *"Gamma sterilization of UHMWPE articular implants: an analysis of the oxidation problem,"* Biomaterials, 17 (18) pp. 1741–1753, 1996.
- Pottage T, Macken S, Giri K, Walker JT, Bennett AM. Low-temperature decontamination with hydrogen peroxide or chlorine dioxide for space applications. *Appl. Environ. Microbial*; 78(12):4169-4174, 2012. doi:10.1128/AEM.07948-11.
- Qiu Q.-Q., Sun W.-Q., Connor J., *"Sterilization of Biomaterials of Synthetic and Biological Origin, in: P. Ducheyne (Ed.), Comprehensive Biomaterials,"* 1st ed., Elsevier Ltd., Amsterdam, pp. 127–144, 2011.
- Rutala W.A., Weber D.J., (HICPAC) Healthcare Infection Control Practices Advisory Committee, *Guideline for Disinfection and Sterilization in Healthcare Facilities, 2008,* Centers for Disease Control and Prevention, 2008.
- Shelar Gaurav A, Gaikwad Sandip T, *"Extrusion in food processing: An overview,"* The Pharma Innovation Journal; 8(2), pp. 562-568, 2019.
- Silindir Mine, Özer A. Yekta, *"Sterilization Methods and the Comparison of E-Beam Sterilization with Gamma Radiation Sterilization,"* FABAD J. Pharm. Sci., 34, 43–53, 2009.
- Smelt JPPM, Hellemons JC, Wouters PC., van Gerwen SJC., *"Physiological and mathematical aspects in setting criteria for decontamination of foods by physical means,"* Int. J. Food. Microbiology, 78, pp. 57-77, 2002.
- *Sterilization methods overview gamma, e-beam, eto, steam & dry heat validations, Life Science Outsourcing,* Inc.Available: http://www.lso-inc.com/sterilizationvalidation- services/sterilization-methods-overview.html , Accessed date: 30 May 2017(Online).

- Soares Goncalo C., Learmonth David A., Vallejo Mariana C., Davila Sara Perez, Gonzalez Pio, Sousa Rui A., Oliveira Ana L., "Supercritical CO2 technology: The next standard sterilization technique?," Materials Science & Engineering, C 99, pp.520–540, 2019.
- USA-FDA Center for Food Safety and Applied Nutrition, 2011.
- White A., Burns D., Christensen T.W., "Effective terminal sterilization using supercritical carbon dioxide," J. Biotechnol. 123 (4) 504–515, 2006.
- Yoganarasimha S, Trahan WR, Best AM, Bowlin, G. L., Kitten, T. O., Moon, P. C., Madurantakam, P. A., *"Peracetic Acid: A practical agent for sterilizing heat-labile polymeric tissue-engineering scaffolds,"* Tissue engineering. Part C, Methods, 20(9), pp. 714–723, 2014.

❒

5

Processing with Supercritical Carbon Dioxide

5.1 Unique Advantages of $SCCO_2$-based-Technologies

After analyzing the demerits and limitations of different conventional sterilization and preservation processes, it is now most appropriate to explore an alternative novel emerging technology using supercritical carbon dioxide ($scCO_2$) for the same purpose taking advantage of its unique properties. $ScCO_2$ is a supercritical fluid (SCF) which is defined by its thermodynamic state when both temperature and pressure exceed their critical point values. As can be seen from *Figure 5.1*, carbon dioxide (CO_2) becomes a supercritical fluid (SCF) above its critical temperature (31.1°C) and critical pressure (73.8bar).

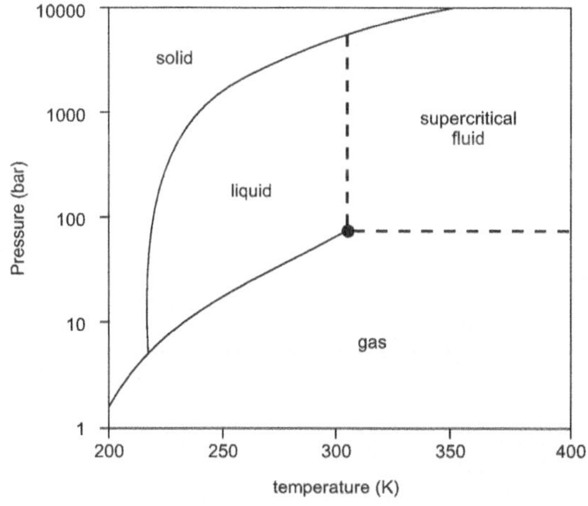

Fig 5.1: Thermodynamic states of carbon dioxide in pressure-temperature phase diagram

M. Mukhopadhayay and A. Chatterjee, *Sterilization and Preservation*,
https://doi.org/10.1007/978-3-031-17370-7_5

Table 5.1 lists critical conditions of different solvents. It can be seen that the critical point of CO_2 is comparatively low and easily attainable. For example, the critical point of water is at T_c = 374.2°C and P_c = 220.5 bar. By far the most popular supercritical fluid is $scCO_2$ owing to its easy availability, low cost, and inherent non-toxicity. The most attractive feature of an SCF is that it has transport properties like those of the gases and density like that of liquids (as given in *Table 5.2*). The solvent power of an SCF solvent is comparable to that of liquids because of its liquid like density. Its diffusivity is comparable to that of gases. The SCF solvent has a negligible surface tension which lies between the vapor and liquid phase values near the critical point. Variations in the properties of fluid with temperature at supercritical conditions are shown in *Figure 5.2*. It can be seen that above the critical point, there is a continuity of all thermo-physical properties of $scCO_2$. The SCF solvent is highly compressible and the change of density with pressure and temperature at the critical point is infinite.

Table 5.1: Critical properties of various solvents *[Reid et al., 1987]*

Solvent	Molecular weight g/mol	Critical temperature K	Critical pressure MPa (atm)	Critical density g/cm^3
Carbon dioxide (CO_2)	44.01	304.1	7.38 (72.8)	0.469
Water (H_2O)	18.02	647.3	22.12 (218.3)	0.348
Methane (CH_4)	16.04	190.4	4.60 (45.4)	0.162
Ethane (C_2H_6)	30.07	305.3	4.87 (48.1)	0.203
Propane (C_3H_8)	44.09	369.8	4.25 (41.9)	0.217
Ethylene (C_2H_4)	28.05	282.4	5.04 (49.7)	0.215
Propylene (C_3H_6)	42.08	364.9	4.60 (45.4)	0.232
Methanol (CH_3OH)	32.04	512.6	8.09 (79.8)	0.272
Ethanol (C_2H_5OH)	46.07	513.9	6.14 (60.6)	0.276
Acetone (C_3H_6O)	58.08	508.1	4.70 (46.4)	0.278

Table 5.2: Density, diffusivity and viscosity for typical liquids, gases and SCFs *[Mukhopadhyay, 2000]*

State	Density (Kg/m^3)	Viscosity (cP)	Diffusivity (mm^2/s)
Gas	1	0.01	1-10
SCF	100-1000	0.05-0.1	0.01-0.1
Liquid	1000	0.5-1.0	0.001

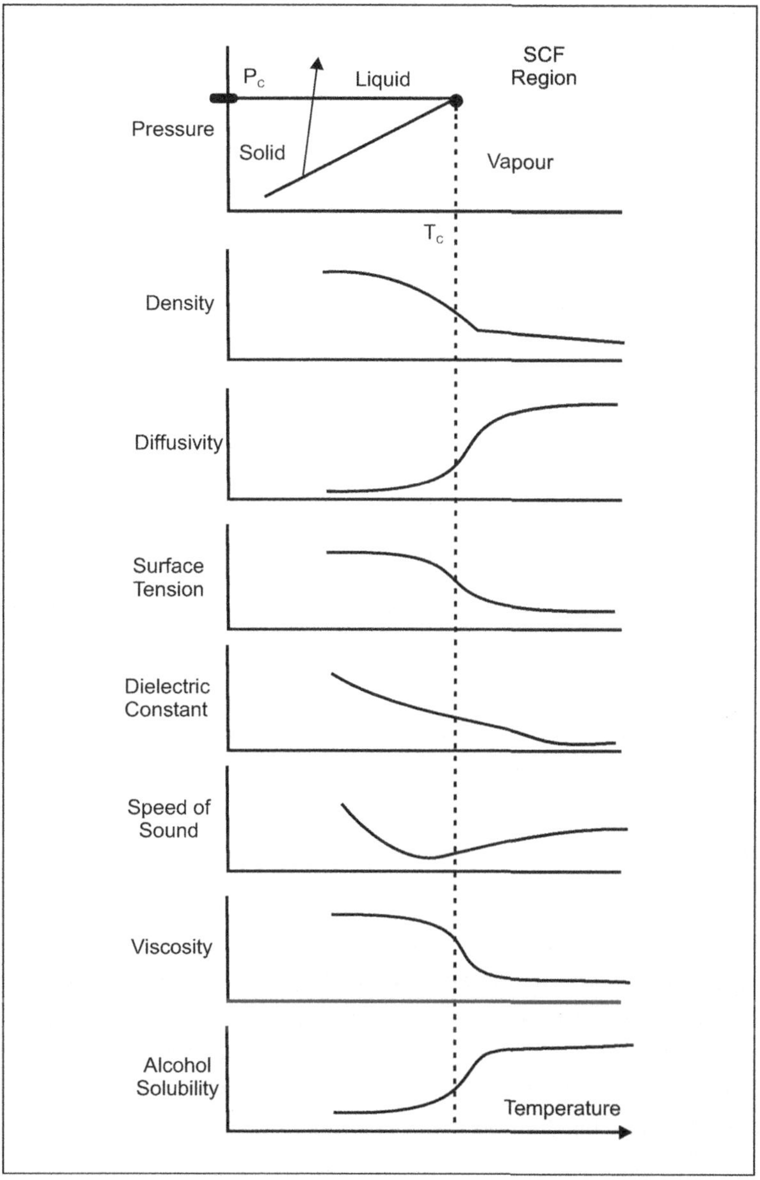

Fig.5.2: Variation of thermo-physical properties of SCF with temperature *[Thiering et al., 2001]*

The SCF solvent has the advantage of having high density (offering the maximum solvent capacity) as well as high compressibility (offering the largest variability of solvent power by small changes in temperature and pressure), as can be seen in *Figure 5.3*. The SCF solvent allows solute to exhibit relatively high diffusivities *[Thiering et al., 2001]*. It offers very unique characteristics owing to its favorable diffusivity, viscosity, surface tension, and other physical properties which make it a unique choice for a variety of processing applications. Finally, both the diffusivity and the density of $scCO_2$ can be tuned by pressure and temperature control thus making $scCO_2$ a versatile medium.

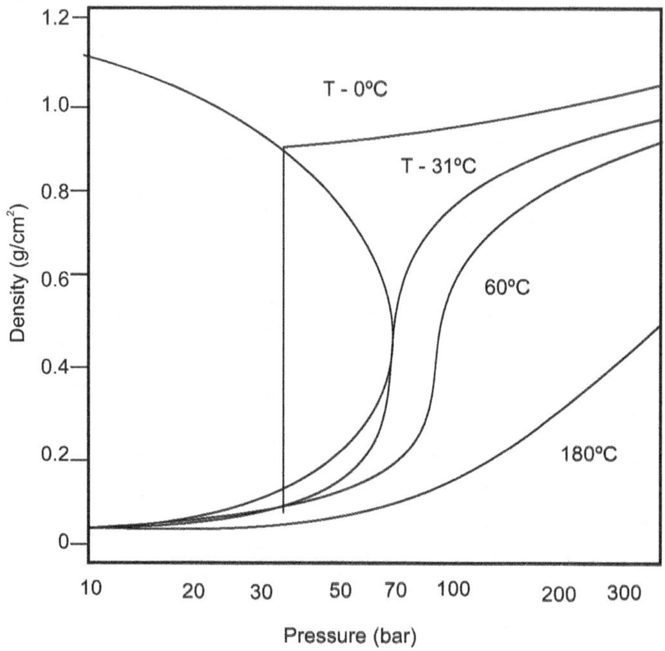

Fig 5.3: Carbon dioxide density variation with pressure and temperature

Dense CO_2 or $scCO_2$ could very well be the most commonly used solvent in this century in wide-ranging applications in processing nutraceuticals, foods, flavors, fragrances, cosmetics, and biologically active principles. Some of the unique features of $scCO_2$ and the underlying principles of various current and potential applications of the environmentally-beneficial $scCO_2$- based technologies will be highlighted in this chapter.

5.2 Characteristics of Supercritical Carbon Dioxide

5.2.1 Path for Attaining Supercritical State

Depending on the external conditions, CO_2 can exist in solid, liquid or gaseous phase, and the transition between the two phases is depicted by the phase boundaries, as represented in *Figure 5.4.* A change of state can occur by varying the pressure and temperature conditions. If one follows the liquid–gas phase boundary, with increasing pressure and temperature, the density of the liquid phase decreases and the density of the vapor increases until the two densities become equal, and the demarcation between the liquid and vapor phases disappears, due to their merging into one phase. This phenomenon appears at a point named "critical point" defined by critical temperature (T_c) and critical pressure (P_c). For CO_2, T_c is 31.1°C and P_c is 73.8 bar. .

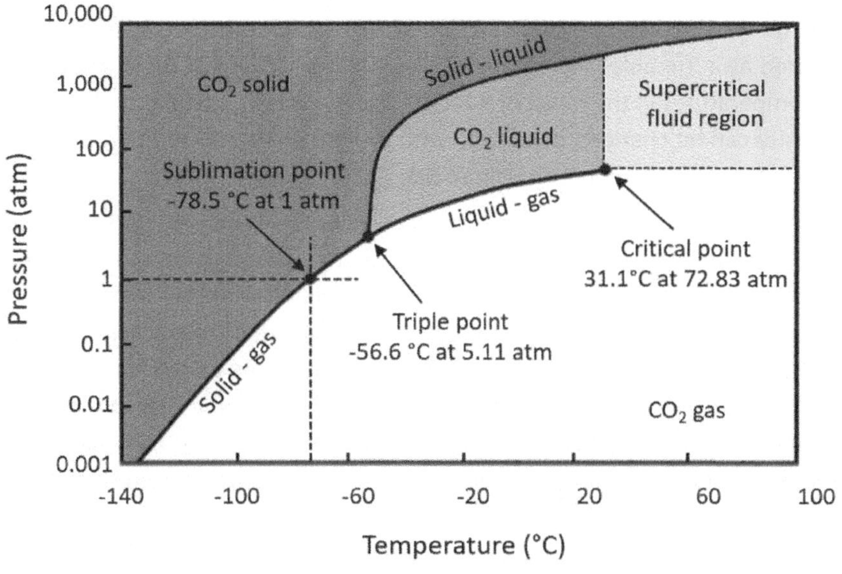

Fig 5.4 : Phase boundaries between solid-liquid, solid-gas, and liquid-gas phases of CO_2

The uniqueness of the critical temperature is that at a temperature above this critical temperature, a substance cannot be liquefied under whatever high the pressure that could be raised to and exerted to it. On the other hand, CO_2 can be easily liquefied under moderate pressures (50-60 bar) at near critical temperatures, requiring negligible cooling duty, as the latent heat is zero at the critical temperature. The liquid CO_2 is then pumped to the supercritical pressure followed by heating to attain the supercritical condition. This path of attaining the supercritical state is preferred, as pumping of liquid CO_2 requires less energy than compression of gaseous CO_2 to attain the supercritical pressure and thus, this route reduces the net energy requirement and also saves high capital cost of compressors.

5.2.2. Unique Advantages and Properties As a Solvent

Carbon dioxide is chemically inert and is easily available in a large quantity with high purity. The physical and transport properties of $scCO_2$ (e.g., density, diffusivity and viscosity) are intermediate between those of the gas and those of the liquid. The density of $scCO_2$ is close to that of a liquid in the range of 0.2- 1.5 g/cm^3 while its transport properties are close to those of a gas. The properties of $scCO_2$ can be tuned to a large extent and continuously from the gas to the liquid properties by varying pressure and/or temperature, unlike those of a liquid solvent. It is thus stated that the properties of several solvents can be made use of in a single $scCO_2$ solvent and thus, $scCO_2$ is generally considered as a 'tunable' solvent. Particularly in the vicinity of the critical point, i.e., at temperatures in the range of 30-40°C or pressures in the range of 75-85 bar, the density can be changed drastically with a slight change in temperature and/ or pressure, as can be seen from *Figure 5.3.*

The solubility of a solid solute in $scCO_2$ increases with pressure or density of $scCO_2$ and decreases with temperature up to a limiting pressure (termed as 'cross-over' pressure), beyond which the solubility increases with both pressure and temperature *[Mukhopadhyay, 2000]*. This phenomenon makes $scCO_2$ a good solvent for extraction of natural substances as will be seen later. CO_2 can be recovered after extraction by simply lowering the pressure or increasing the temperature or both. The additional properties that make $scCO_2$ a most desirable choice for extraction of heat sensitive natural products, food processing and pharmaceutical industries are:

i) It is a gas at ambient condition, critical at 31.1°C, so it is suitable for heat sensitive materials.

ii) It diffuses out of the system, so the problem of residual contamination does not arise.

iii) It is odorless and tasteless, so it does not impart any added flavor to the material.

iv) It removes dissolved oxygen from the system, thus retaining antioxidants in the material.

In recent years supercritical fluid extraction (SCFE) with $scCO_2$ as a solvent has emerged as a highly promising environmentally benign technology for production of natural extracts, such as, flavors, fragrances, spice oils and oleoresins, natural antioxidants, natural colors, nutraceuticals, or herbal medicines with high potency of active ingredients from natural products and for the decaffeination of green coffee beans or to remove contaminants. Worldwide there has been increasing demand for superior quality and safety of foods and medicines, as well as serious concern for environmental protection. These have triggered stringent regulations on the levels of toxins and pollutants. In addition, there has been increasing consumer preference for natural substances. All of these factors have given strong impetus to the development of alternative 'natural' foods and medicines, as well as eco-friendly cost-effective extraction technologies. The need to replace traditionally used organic and aqueous solvents with environmentally benign, non-flammable, and relatively inexpensive alternative solvent has been an important focus of research in academia and industry in the last three decades.

5.3 Supercritical Fluid Extraction Process

The new extraction process uses supercritical carbon dioxide ($scCO_2$) as solvent which is clean, safe, nontoxic, inexpensive, nonflammable, nonpolluting, and environment friendly. It is generally regarded as safe (GRAS) and it yields contaminant-free, tailor-made extracts of superior organoleptic profile and shelf life, with high potency of active ingredients. Its near-ambient critical temperature (31.1°C) makes it ideally suitable for thermolabile natural products. Due to its low temperature of operation the extracts are very close to that in nature, in smell and taste. Dense or supercritical carbon dioxide ($scCO_2$) could very well be the most commonly used solvent in this century, due to its wide-ranging applications in the processing of nutraceuticals, foods, flavors, fragrances, cosmetics, and biologically active principles *[Mukhopadhyay, 2000]*.

SCFE is a two-step process involving extraction and separation by varying the solvent power of the same solvent at different conditions, as schematically described in *Figure 5.5*. The feed, generally ground solid, is charged into the extractor and $scCO_2$ is fed to the extractor through a high pressure liquid CO_2 pump (at 100 - 500 bar). The extract-laden $scCO_2$ is sent to a separator (at 60 - 120 bar) via a pressure reduction valve. At reduced temperature and pressure conditions, the extract precipitates out in the separator due to reduction in its solvent power. However, it is possible to reduce the solvent power of the extractant in the separator, not only by reducing the pressure, but also by increasing the temperature or by addition of a third substance, depending on the nature of the feed and process economics. For example, for coffee decaffeination, either hot water is pumped to the separator or the extract-laden CO_2 is passed through a bed of activated charcoal to remove caffeine from the CO_2 stream. The extract-free CO_2 stream, leaving the separator, is then recycled to the extractor. In the case of liquid feed, the extractor is modified into a column through which feed and $scCO_2$ are fed either co-currently or counter currently *[Mukhopadhyay, 2000]*.

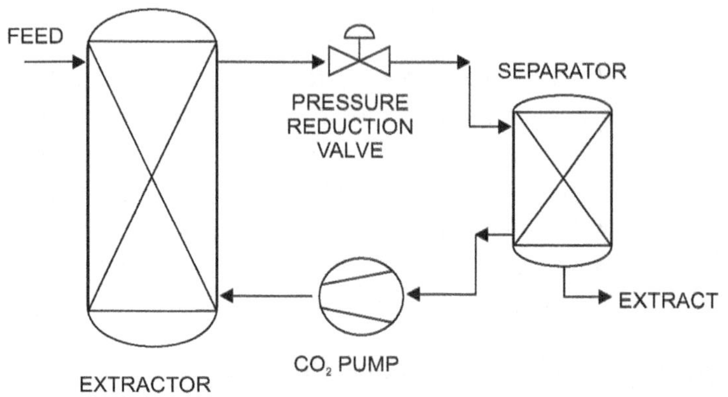

Fig. 5.5: Simple schematics of the principle of SCFE process

Dense carbon dioxide tends to be selective towards low molecular weight compounds, such as, hydrocarbons, halocarbons, ethers, esters, ketones, and aldehydes. It has zero dipole moment and a positive quadruple moment at high pressures. It can selectively extract nonpolar components and moderately polar compounds at higher pressures. One limitation $scCO_2$ is debated to have, is that it is essentially non-polar, i.e., unsuitable for processing water-soluble substances. However, this seeming drawback may be easily overcome by adding a polar co-solvent (typically in very small quantities, say 5 mole %) to $scCO_2$ to increase its polarity and hence the solvent power for highly polar or high molecular weight compounds *[Mukhopadhyay, 2000]*. The binary homogeneous mixture is then capable of extracting water-soluble or high molecular weight compounds. The best candidate for such a co-solvent, especially for foods and nutraceuticals, is ethanol or ethyl acetate or possibly water which is the best natural food-grade cosolvent.

One of the major advantages of $scCO_2$ as a solvent is the possibility to recover the final product in a dry, free of any residual organic solvents, and in sterile condition, since CO_2 is a gas at ambient pressure and can be easily removed by depressurization. This property is interesting from an economical point of view, since it avoids the many purification steps that are usually performed after processing with toxic organic solvents. Moreover, the recovered CO_2 can be easily separated from other compounds, such as, Active Pharmaceutical Ingredients (APIs) for recycling.

In view of India's rich botanical and marine resources, SCFE has high potential in producing natural concentrates of flavors, fragrances and biologically active principles, and thereby achieving a significant value-addition to its traditional export of raw natural materials. Thus use of the SCFE technology can help Indian industry to gain its share in rapidly growing international market for high quality value-added natural extracts.

For example, the SCFE technology can be profitably utilized for production of phytochemicals and nutraceuticals *[Martinez,2008]*. Phytochemicals are naturally occurring phenolic or polyphenolic compounds present in leaves, roots, stems, seeds, and fruits. Some of them are important natural antioxidants which deter the formation and propagation of free radicals. These antioxidants are added in small concentrations to foods as supplements for preservation and enhancing nutrients, e.g. vitamins A, C, E, and flavonoids. Nutraceuticals (often referred to as phytochemicals or functional foods) are natural bio-active chemical compounds that have health promoting and disease preventing medicinal properties. These health supplements are obtained from natural products. The market for nutraceuticals is now steadily growing in all four sectors, namely, (a) the food industry, (b) the herbal and dietary supplements industry, (c) the pharmaceutical industry, and (d) the newly merged pharmaceutical / agri-business / nutrition conglomerates *[Mukhopadhyay, 2009]*. The greatest scientific need in nutraceuticals industry now entails high consistency and reliability in the quality and safety of the final product. These are achievable by adopting the SCFE technique, as these bio-active heat-sensitive natural molecules can be recovered in natural form without degradation and contamination. Extraction with $scCO_2$ at low temperatures does not alter the delicate balance of bio-activity of natural molecules, making SCFE a superior alternative to the conventional steam distillation (SD) and solvent extraction (SE) techniques in food, pharmaceutical, and nutraceutical industries *[Mukhopadhyay, 2008]*. Processes to obtain vitamin additives, de-alcoholized, beverages, de-fatted potato chips, and encapsulated liquids using this technology are well established. Some examples of food products consumed in day-to-day life and produced by SCFE technology are listed in *Figure 5.6*.

Fig 5.6: Supercritical fluid technology applied to everyday food *[Brunner, 2005]*

5.4 Commercial Scale Supercritical Fluid Extraction Process

Supercritical fluid extraction (SCFE) in commercial scale is usually a semi-batch process in which CO_2 flows in a continuous mode, whereas the feed is charged in the extractor basket in batches. However, for better viability of the commercial scale process, its operation is made semi-continuous using multiple extraction and separation vessels, positioned sequentially. The extraction and separation of the extract are often carried out in stages, by sequentially maintaining different conditions of pressure and / or temperature in the extractors and separators. This allows easy fractionation of the extract for enrichment of the specific active components, which are subsequently precipitated in each of the separators. In essence, it is possible to avail of a spectrum of solvent properties in a single solvent by merely changing the pressure, temperature, and co-solvent concentration *[Mukhopadhyay, 2000]*, often required for a variety of products using the same hardware, facilitating it to be a truly multi-product plant. A number of innovations *[Mukhopadhyay and Sastry, 1996; Mukhopadhyay and Kalyan Ram, 1997, and Mukhopadhyay and Karamta, 2004]* have been developed at IIT Bombay, incorporating which the conventional SCFE processes can be further improved, as the yield and efficiency of extraction can be further enhanced and the extraction pressures may be lowered.

The commercial scale SCFE plant has three extractors, two separators, one CO_2 hold-up tank and heat exchanger, as shown in the flow diagram *(Figure 5.7)* of a commercial prototype pilot plant indigenously developed and built at IIT Bombay for the purpose of SCFE technology development and transfer *[Mukhopadhyay, 2000]*. The extraction process works on semi batch mode of operation with a typical batch time of 2 to 4 hours in a closed loop of constant circulation of CO_2 in the system. CO_2 is liquefied and fed to the hold-up tank through a heat exchanger (condenser). Liquid CO_2 is then fed to the high-pressure metering pump, through a heat exchanger (sub-cooler) to ensure supply of CO_2 in liquid form. The high-pressure liquid from the pump is then heated in a heat exchanger (pre-heater) to achieve supercritical temperature.

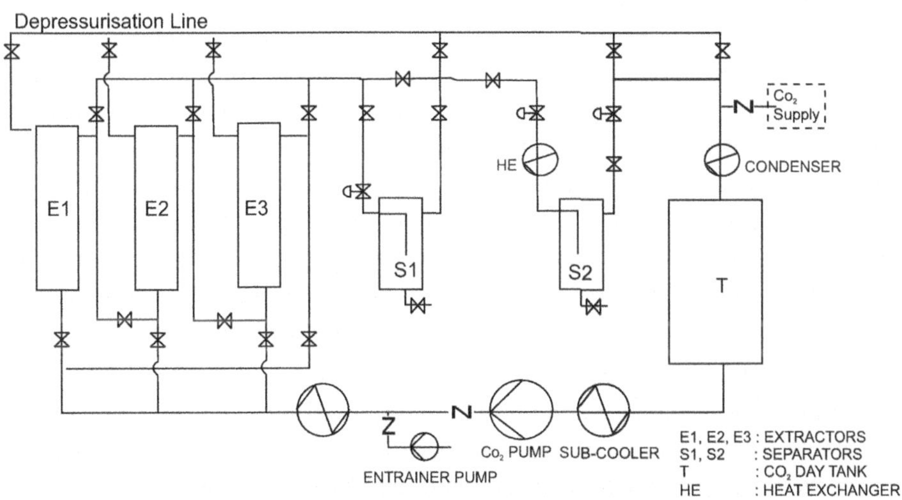

Fig 5.7: Schematics of a commercial protoype SCFE pilot plant developed at IIT Bombay

A co-solvent pump can be used for dosing a small quantity of other solvents, such as, water, ethanol, etc., into the CO_2 stream, if desired. This fluid mixture then flows through the extractors. The $scCO_2$ with dissolved extract then flows to separators. The CO_2 vapor leaving the last separator is condensed and recycled to CO_2 hold-up tank. When the solid feed in the extractor needs to be removed or when plant is to be shut down, the respective vessels are depressurized to atmospheric pressure. While doing so, the maximum possible CO_2 is fed back to the CO_2 hold-up tank. The remaining CO_2 is then vented off to the atmosphere *[Mukhopadhyay, 2000]*.

Typical operating pressure and temperature needed for this application are in the range of 100 - 500 bar and 40 - 80^{0}C. There is a common apprehension that the SCFE technology is energy-intensive due to the requirements of the high pressure-operation and very accurate process control. However, it is important to note that the energy needed for attaining the supercritical state (P_c = 73.8 bar) is more than compensated by the negligible energy required for the solvent recovery from the extract, which is a simple depressurization step for precipitation of the extract and recovery/recycle of CO_2 . As a result, the overall energy expended is lower than that which is characteristic of traditional SD or SE. The relatively higher capital investment required in the SCFE process is well compensated by the reduced operating costs due to other benefits of the process, such as, low solvent (CO_2) cost, lower batch times, higher concentration of active / desirable components in the extract, no pollution control related costs, etc. There is practically no effluent generation in SCFE. In addition, the extracted residue (cake) from the SCFE process does not undergo any degradation, unlike in SE and SD. Thus it retains all the useful ingredients, such as, edible proteins and fibers, etc. This residue can be sold as a high value by-product to give additional revenue to the project. In addition, the accurate control over the SCFE process which can be achieved with state-of-the-art instrumentation and control system ensures high consistency in the high quality of the final product. This is an important factor in favor of commercial scCO_2 extracts for their preference in the global market. A detailed feasibility study shows that even at the existing price (of extracts from SE and SD) of oil and oleoresins, the investment on SCFE is profitable which justifies it to be the preferred route on long term perspective *[Mukhopadhyay 2000]*.

The efficacy and versatility of SCFE is well recognized today and its utilization brings economic and social benefits. In essence, SCFE is economically profitable for an appropriate combination of high-value low-volume multiple products. Larger the capacity of the plant, higher is the return on investment, despite the fact that initial investment increases. For example, in a semi-continuous commercial coffee decaffeination plant, there may be a minimum of four extractors each having an optimal capacity of as high as tens of cubic meters with a height of 23 meters, whereas, for commercial production of high-value herbal extracts, the optimal capacity of an extractor may be as low as one hundredth of a cubic meter. The Maxwell House Division of General Foods has employed supercritical extraction process for production of 25K MTA decaffeination coffee in Houston, USA *[Mukhpadhyay 2000]*.

5.5 Commercial Applications of SCF Extraction from Natural Products

SCFE has successfully and totally replaced chlorinated solvents in applications like decaffeination of tea and coffee, hops extraction, denicotinization of tobacco, etc. Similar trend is observed in other applications using organic solvents as well. The trend is gaining momentum and thus solvent extraction process may be discontinued. Thus the industry needs to change over to this modern technology which can improve the quality and reliability of the spice extracts, herbal medicines, and nutraceuticals. The present international trend of increasingly stringent regulations regarding residual organic solvents in processed food product clearly indicates that the conventional solvent extraction process may be phased out in the near future. Already chlorinated solvents are banned from their use in food processing.

The remarkable value-addition that the $scCO_2$ extracts offer as natural concentrates, in addition to the advantages from the points of view of environment and health, has generated a great deal of commercial interests in this technology. It is a well established fact that the nutraceuticals produced by SCFE are preferred by consumers due to their superior quality and bio-activity and higher yields without the problems of residual solvent and microbial contamination. Due to the relatively low temperature of operation, the extracts are free of artifacts and like natural. These extracts have longer shelf life due to co-extraction of antioxidants. Some of the commercial areas and products for which supercritical fluid technology has been successfully developed and used are given in *Tables 5.3 to 5.5*.

Table 5.3: Commercial products by $scCO_2$ extraction *[Mukhopadhyay, 2000]*

- Decaffeinated (Coffee and Tea)
- HOP Extracts (Bitter)
- Spice Extracts (Oil & Oleoresin)
- Flavors & Fragrances
- Food Colors
- Food Preservatives
- Herbal Medicines
- Pesticides (Neem)
- Deoiling of Fast Foods
- Cholesterol – Free Food Products
- Nicotine / Tar free Tobacco
- Natural food Colors, Anthocyanins
- Antioxidants (Vitamins E and A, Flavonoids, Carotenoids
- Nutraceuticals, Alkaloids, Polyphenols

Table 5.4: Some commercially available natural extracts *[Mukhopadhyay, 2008]*

Black Pepper	Marjoram
Cardamom	Nutmeg
Celery	Paprika
Clove	Pimento
Ginger	Rosemary
Garlic	Sage
Hops	Vetiver
Mace	Vanilla

Table 5.5: Application Areas of $scCO_2$ Technology for Vegetable / Animal Oil Processing *[Mukhopadhyay, 2000]*

i) Separation of FFA from Vegetable Oils
ii) Fractionation of PUFA from Animal Lipids
iii) Refining and Deodorization of Vegetable Oils
iv) Fractionation of Glycerides
v) Recovery of Antioxidants (Vit. E and A)
vi) Extraction of Oil and ω-3 fatty acids from Oil-bearing Materials
vii) Deoiling / Purification of Lecithin
viii) Decholesterolization of Butter, Egg, Fish, Meat muscles
ix) Deoiling of Snack-foods
x) Recovery of GLA (evening Primrose seed oil, Borage seed oil), MUFA (Tea seed oil)
xi) Recovery of ω-3 fatty acids (Marine Fish Oils)

5.6 Other $SCCO_2$- based Processes

The unique advantages of $scCO_2$-based processes have led to the development of a variety of new products and production of novel materials with wide-ranging applications. In addition to the unique properties of $scCO_2$, the advantages like generation of minimum wastes and residual-solvent free products, easy separation of solutes and recycling of solvent, a solvent medium for fast and selective reactions, etc., have given impetus to development of new $scCO_2$–based processes for chemical reactions, metal-ion separations, environmental remediation, reprocessing of spent nuclear fuels, nuclear waste management, sterilization and preservation of food products and biomaterials, clinical waste management, and pollution abatement *[Mukhopadhyay, 2009]*. Environmental concerns, regulations, and cost effectiveness are driving the industry's proactive stance in the development of environmentally benign processes. The newly developing paradigm of "industrial ecology" recognizes the importance of the coexistence of both the world's ecosystem and the industrial system. Industry has begun to understand that an improvement aimed at pollution prevention, commonly called green chemistry and green engineering, are most cost-effective. Nowadays, the development of sustained chemical processes has become a priority for the chemical industry because of environmental concerns and of stricter legislation. In this context with a view to meeting these criteria, $scCO_2$- based technologies have emerged as a good alternative to replace several conventional processes *[Champeau et al., 2015]*.

Currently the $scCO_2$-assisted micronisation, micro-encapsulation, and impregnation processes are commercially used in pharmaceutical and cosmetics industries for production of sub-micron particulates of uniform size and controlled morphology *[York et al., 2004]*. These are needed for efficient drug delivery applications and particularly for controlled release of the biologically active compounds that render higher dissolution rates and better bio-availability in the body fluids. The $scCO_2$ technology has also been successfully used for sterilization and stabilization of food products to enhance shelf life and safety, with no additives, degradation, and contamination *[Mukhopadhyay and Chakraborty, 2004]*.

5.6.1 Micronization, Microencapsulation, and Impregnation for Drug Delivery

In recent times, a variety of medical devices, such as, dry powder inhalers, needle-free injections, micro-encapsulated drugs are adopted for controlled release. These require, micronized drugs with a narrow size distribution in the range of few nanometers (nm) to few microns (μm). These are more effective for absorption by the target organs like the heart, the lung, the tissues, the bones, etc., as they perform better than those with larger sizes and broader size distribution. The $scCO_2$-based micronization process is in demand, as it produces particulates that are free from any organic solvent residues and artifacts. This process is preferred as it involves single-step operation, mild operating temperature, and produces very narrow size distribution of sub-micron particles with controlled morphology. It employs mild processing conditions and provides excellent properties enabling formation of micro-particles, without the problem of residual solvent. On the contrary, the conventional micronization processes, like co-acervation (phase separation), spray drying, micro-emulsion and evaporation techniques, ball milling and sieving, pneumatic jet grinding, etc., have a number of disadvantages. These are multi-step processes and require large quantities of organic solvents and consequent complex recovery and purification steps, poor control of particle size and size distribution, and thermal degradation reducing the bio-activity of the drug *[York et al.,2004]*.

The $scCO_2$-based micronisation processes, in general, involve dissolution of the active pharmaceutical ingredient (API) in a suitable organic solvent or in $scCO_2$ itself as the solvent, followed by rapidly lowering its solubility respectively by adding CO_2 as antisolvent or by reducing the pressure over the solution , resulting extremely rapid particle formation *[Thiering et al, 2001]*. When $scCO_2$ is used as the solvent, the process is known as in Rapid Expansion of Supercritical Solution (RESS) or Particles from Gas Saturated Solutions (PGSS), depending on the state of the solution before depressurization, as the name suggests. When an organic solvent is used as the solvent, then the process is known as Gaseous Anti-Solvent (GAS) or Supercritical (CO_2) Anti-Solvent (SAS), depending on whether CO_2 is in subcritical gaseous or supercritical state *[Mukhopadhyay and Dalvi, 2004]*. A similar process, commonly known as the Solution Enhanced Dispersion by Supercritical fluid (SEDS) also utilizes $scCO_2$ mixed with a cosolvent as the antisolvent to produce solvent-free drug microparticles. $ScCO_2$ is highly soluble in an organic solvent and vice versa.. The GAS process can also be used for purification of pharmaceutical substances by dissolving the solid impure drug in an organic solvent in which CO_2 is highly soluble. By passing CO_2 into their dilute solutions, the solid impurities are selectively separated in a sequence of their decreasing polarity. These processes using $scCO_2$ as the processing medium have been found promising for large scale commercial applications in pharmaceutical industries *[York et al.,2004]*.

The drug delivery system for controlled release uses the process of micro-encapsulation of drugs for coating of the biologically active compounds. The controlled drug release is used for treatment of diseases ranging from diabetes to cancer. The principal advantage of these drug delivery devices is that the total amount of effective drug used is reduced, yet its effect is prolonged avoiding the problems associated with repetitive introduction of excessive dosages. There are two important methods for controlling the release of drugs, namely, by impregnation of the drug in a polymeric matrix or by micro-encapsulation of the drug in the core. The polymeric matrix used for controlled release of drugs needs to be biocompatible or biodegradable and nontoxic to the body system. Such biocompatible polymeric matrices could be natural polymers (such as, collagen, cellulose, and chitosan) and recently investigated synthetic polymers (such as, polyester, polyurethanes, and polyanhydrides). More commonly used biodegradable polymers are PLA, PHB, PDLA, and PGLA, and are produced as micro spheres using $scCO_2$ –based processes *[York et al.,2004]*.

For example, monodisperse polymer globules (of the order of 10 microns) encapsulating hydrophilic drugs, can be produced using RESS or PGSS using $scCO_2$. Polymeric microparticles encapsulating flavonoid, such as, 3-hydroxyflavone as the core material, are produced from a suspension of flavonoids in $scCO_2$ mixed with a co-solvent and a dissolved polymer (e.g., aminoalkyl methacrylic copolymer) by spraying it through a capillary to atmospheric pressure. Small amounts of cosovlents, such as, alcohols are used for increasing the solubility of polymer in $scCO_2$ The microparticles thus produced do not agglomerate, though flavonoid is coated thoroughly by the polymer. The particle size and size distribution of microcapsules can be controlled by changing the copolymer feed composition. The drug delivery system is designed such that the release rate of flavonoids from microcapsules is controlled by pH of the body fluid.

Another example for controlled drug delivery system is protein-encapsulated polymeric nanoparticles using $scCO_2$ based micro-encapsulation process. In order to retain the biological activity of precipitated proteins, they are entrapped into microparticles of bio-polymer matrix. The biodegradable polymer, poly-l lactide acid (PLA) microparticles are produced by spraying the solution of the protein and the polymer in an organic solvent into $scCO_2$ medium, precipitating both, while encapsulating protein in polymer. Similarly lysozyme or insulin is encapsulated in polymer microparticles with a diameter ranging between 1-5 μm by spraying a homogeneous protein and polymer organic solution using the $scCO_2$-based SAS process *[York et al., 2004]*.

In recent years, for controlled drug delivery applications, the $scCO_2$-assisted impregnation process has been employed for the preparation of polymeric implants as medical devices for maintaining prolonged effect of the drug by sustaining its release. This requires a swelling agent for the polymer which is also a solvent for the drug, in order to introduce the drug into the polymer matrix for impregnation. $ScCO_2$ has been used as the swelling agent as well as a solvent for impregnation of the polymer matrix with the drug. $ScCO_2$ is known to reduce the glass transition temperature (T_g) for many polymers and the degree of swelling can be manipulated by changing the pressure. As CO_2 is a smaller molecule, it diffuses more quickly than conventional liquid solvents in polymers, so that the swelling is more rapid. Swelling facilitates the diffusion of a solute drug, increasing the rate of impregnation. CO_2 is released very rapidly and fully from the polymer on depressurization, and the drug is precipitated on the external surface or impregnated inside the polymer beads *[McHugh, 1986]*.

5.6.2 Synthesis of Novel Materials, Specialty Chemicals, and Polymers in $ScCO_2$-Medium

One remarkable recent development in the use of SCF fluids is in the synthesis of novel materials and specialty chemicals. This is based on the principle that the solubilizing capacity of the SCF solvent can be fine- tuned easily with the density of the SCF solvent by variations of pressure and temperature. The large capital cost of the pressurized reactor can be balanced by the savings from combining a reaction and a separation into a single process unit. The same variations can also be utilized to optimize the reaction rates, selectivity, and yields of reactions in an SCF medium. The significant advantages that it offers is that all fluid-phase reactants may be combined into a single homogeneous phase, thus minimizing gas-liquid mass transfer limitations to a higher conversion *[Mukhopadhyay, 2009]*.

Owing to its essentially benign nature, $scCO_2$ is among the most frequently used truly green media for reactions as well. For example, it offers a chemically inert atmosphere for many heterogeneous conversions using zeolites as catalysts. The main advantages in such an $scCO_2$-based process are that with increasing pressures in the supercritical condition, the catalyst activity is enhanced and catalyst deactivation is strongly reduced at the same time. Reactions between immiscible species are often carried out with a phase transfer catalyst (PTC). These PTCs are generally soluble in non-polar solvents. The ease of removal of $scCO_2$ (as well as the tunability of its properties) makes it an attractive solvent for PTC systems. For example, the reaction of benzyl chloride with KBr in presence of tetra heptyl ammonium bromide, can be carried out in $scCO_2$ to get benzyl bromide *[Mukhopadhyay, 2009]*.

Good solubility of most monomers and poor solubility of polymers in $scCO_2$ makes it an ideal medium in the synthesis and processing of polymers. $ScCO_2$ has been widely used for dispersion and precipitation polymerization. The polymer, instead of forming aggregated polymer precipitates, forms spherical particles, in presence of a polymeric stabilizer from an initially homogeneous reaction mixture of monomer, initiator, stabiliser, and solvent. For many macromolecules, $scCO_2$ may be a poor solvent which can be overcome by the addition of certain stabilizers to help emulsify insoluble hydrophilic or lipophilic substances in $scCO_2$. For example, two types of emulsifiers: (1) Perfluoropolyether (PFPE) and (2) Siloxane based surfactants have been reported to provide high solubility for otherwise insoluble substances in $scCO_2$ medium, through the formation of microemulsions *[Mukhopadhyay, 2009]*.

5.6.3 Biochemical Reactions in $ScCO_2$

Several useful enzymes are used as biocatalysts in $scCO_2$ medium for biochemical reactions. Among the enzymatic reactions studied in $scCO_2$ medium so far, the use of *lipase* seems to be most promising for commercial use. It has been well known that *lipase* has been widely used as catalysts for several such reactions (e.g., modification of edible oils and fats, resolution of optical isomers, and synthesis of several esters). The *lipase*-catalyzed reactions are employed in $scCO_2$ medium to provide triglycerides with the attached acyl groups in the specific positions for attaining the desired physical properties or degree of unsaturation. These reactions carried out in $scCO_2$ environment reveal that the enzyme shows very good stability, activity, and specificity in $scCO_2$. The stability of enzymes in $scCO_2$ however, depends on temperature, moisture content, pressurization-depressurization sequence, etc. However the enzymes lose catalytic activity in $scCO_2$ if the moisture level in $scCO_2$ medium increases beyond a certain critical value. It is known that the water solubility in $scCO_2$ may be in the range of 0.3-0.4 wt% water depending on pressure and temperature, thus maintaining the required level of water bound to enzymes during the progress of the enzymatic reactions which produce water as the byproduct. *Lipase* remains active over $2.0 < pH < 8.5$. As mentioned earlier, $scCO_2$ as the reaction medium offers specific advantages over the organic solvents, due to its low toxicity, easily controllable solubility of components, high diffusivity, and improved reaction rates. As a result, the time required for enzymatic esterification of fatty acids from milk fat is approximately 1 hour in $scCO_2$ medium, as compared to 120 hours required in an organic solvent, such as, ethanol *[Mukhopahyay, 2009].*

Another important biological application of $scCO_2$ is in sterilization and preservation of food products by selective inactivation of certain enzymes, such as, *pectin esterase* and *polyphenol oxidase*. The decay of enzyme activity depends on the water content, pH value, pressure, and temperature. The wet microbial cells and spores are also disrupted during the depressurization step, thus enabling selective sterilization by $scCO_2$, which will be elaborated in subsequent chapters.

5.7 Conclusion

Currently $scCO_2$ technologies are of great interest. The unique properties of $scCO_2$ facilitate recovery of naturally occurring bio-molecules without degradation and contamination in their natural form. It has been largely used in the foods, beverages, confectionary, cosmetics, and pharmaceutical industries. It is believed that herbal extracts with assured quality have better efficacy than the isolated active ingredients due to their naturally occurring synergistic effects *[Mukhopadhyay, 2000]*. Vast botanical and marine resources available in India can be utilized in producing herbal medicines, nutraceuticals, and bio-active principles with SCFE Technology for healthcare and for value-addition of natural products instead of exporting the raw materials *[Mukhopadhyay and Karamta, 2008]*.

In recent years, enormous progress has been made in developing $scCO_2$-based technologies towards establishing $scCO_2$ as solvent, antisolvent, and reaction medium for a variety of applications. $ScCO_2$ provides an excellent medium for a host of physical processes, chemical, biological and polymerization reactions, and other viable and commercially attractive applications. The stage is now set to adopt $scCO_2$ as the sterilization and preservation agent for foods, biomaterials and medical devices, and for management of clinical waste for safety and protection of the people and environment.

References:

- Brunner, G.; *Supercritical fluids: technology and applications to food processing, J.* Food Engineering, 67, 21, 2005.
- Champeau, M., Thomassin, J.-M., Tassaing, T. .Jérôme C, *"Drug loading of polymer implants by supercritical CO_2 assisted impregnation: A review"*, Journal of Controlled Release 209 (2015) 248–259 249.
- Martinez, J.L, Editor, *"Supercritical Fluid Extraction of Nutraceuticals and Bioactive Compounds,"* CRC Press, 2008.
- Mc Hugh, M.A., *"Supercritical Fluid Extraction, Second Edition, Butterworth-Heinemann Series in Chemical Engineering"*, 1986.
- Mukhopadhyay, M., *"Processing of Spices Using Supercritical Fluids"*, Chapter 11, pp.337 in *"Supercritical Fluid Extraction of Nutraceuticals and Bioactive Compounds"*, Editor Jose L. Martinez, CRC Press, 2008.
- Mukhopadhyay, M., *Natural Extracts Using Supercritical Carbon dioxide* (2000), CRC Press, Florida, U.S.A.
- Mukhopadhyay, M and Sastry, S.V.G.K., *"Process for Supercritical Fluid CO_2 Extraction of Fragrances (absolute or essential oils) from Jasmine flowers,"* Indian Patent No.183454 dated 01.08.2000, Application No. (72/Bom/96).
- Mukhopadhyay, M. and Kalyan Ram, T.V., *"Process for Sequential Supercritical CO_2 Extraction and Fractionation of Neem Oil Enriched with Azadirachtin from Neem kernels"*, Indian Patent No.182587 dated 12.03.1999, Application No. 428/BOM/97.
- Mukhopadhyay, M. and Karamta,H., *"A Novel Process for Nutraceutical Concentrate using Supercritical Fluid Extraction"*, Indian Patent No.213518 dated 01.07.2008, Application No.545/MUM/2004.

- Mukhopadhyay Mamata and Chakraborty Anuradha, *"Process for Sterilization of Biomaterials Using Supercritical Fluids",* Indian Patent No.282732, dated 24.04. 2017 Application No. 543/MUM/2004.
- Mukhopadhyay, M., *"Extraction and processing with supercritical fluids",* Journal of Chemical Technology and Biotechnology, Volume: 84 Issue: 1, pp. 6-12, Jan 2009.
- Mukhopadhyay, M , Dalvi, SV, *"Mass and Heat Transfer Analysis of SAS: Effects of Thermodynamic States and Flow Rates on Droplet Size",* The Journal of Supercritical Fluids, Volume 30, Issue 3, pp. 333-348, August 2004.
- Mukhopadhyay, M. and Karamata, H. *"A Novel Process for Supercritical Fluid Extraction of Nutraceuticals Enriched with Carotenoids"* Indian Chemical Engineer, Vol.50, No.2, 2008, pp.116-121.
- Reid, R C, Prausnitz, J M, and Poling, B E., *The properties of gases and liquids.* McGraw-Hill Education Publishers, NY, 1987.
- Thiering, Russell Dehghani, Fariba, Foster, Neil R, *"Current issues relating to anti-solvent micronisation techniques and their extension to industrial scales",* The Journal of Supercritical Fluids, Volume 21, Issue 2, October 2001, pp. 159-177.
- York, P., Shekunov, B.Y., Kompella, U.; *Drug Delivery and Supercritical Fluid Technology,* Marcel Dekkar, New York, U.S.A.

❐

6

Sterilization and Preservation of Solid Foods with Supercritical CO_2

Food preservation is a process that is undertaken to maintain foods with the desired properties for the longest possible duration. The entire food production, preservation, and packaging industry in the recent years is centered around food safety, incorporating innovation and sustainability. Compromising certain qualities to an extent, such as, color of the food is acceptable; however, where the safety aspect of food is concerned no compromise is allowed. Thus, food safety parameters are stringent and closely watched. After storage for a certain period, one or more quality attributes of a food may reach an undesirable state, leading to a change such that the characteristics of food may reach an unsuitable state for consumption. This is when it is said to have reached the end of its shelf life. To conduct a study on shelf life of foods, it is critical to measure the rate of change of a given food characteristic or quality attribute. The quality of a product may be defined using many factors which include appearance, yield, organoleptic characteristics, enzyme, and microbial characteristics. This chapter presents preservation, sterilization, and stabilization of some of the solid food products using supercritical CO_2 ($scCO_2$).

M. Mukhopadhayay and A. Chatterjee, *Sterilization and Preservation*,
https://doi.org/10.1007/978-3-031-17370-7_6

6.1 Performance Evaluation of $scCO_2$ for Sterilization and Stabilization

A novel process using $scCO_2$ was established at IIT Bombay for sterilization and stabilization of several biological and food systems, varying in moisture, pH, and physical state *[Mukhopadhyay and Chakraborty, 2004]*. The mechanism of the process was first developed (as will be presented later in Chapter 8) by systematically performing experimental parametric studies on microbial inactivation of pure and mixed cultures. Subsequently the same process conditions were applied to the selected food systems for validating the process and ascertaining the optimum process conditions. The cultures selected were *Sacharomyces cerevisiae, Escherichia coli, Bacillus subtilis,* and *Lactobacillus rhamnosus*, since these are the most commonly found organisms in foods. The solid food systems selected for sterilization and preservation experiments were rice bran and soybeans, and the enzymes considered for stabilization experiments were *pectinesterase, lipoxygenase, trypsin inhibitor,* and *lipase [Mukhopadhyay and Chakraborty, 2005]*. This section will describe the experimental process adopted for sterilization and stabilization of rice bran and soybean, followed by performance of this novel process for these two selected solid products for its validation in order that it can be applied to other solid food products.

6.1.1 Sterilization and Stabilization of Rice Bran

Rice bran is a by-product obtained from the outer layer of the brown-husked rice kernel during milling to produce white rice. It is rich in nutrients with 14-16% protein, 12-23% fat, and 8-10% crude fiber. It is also a good source of vitamin B complex and also contains minerals, such as, iron, potassium, calcium, chlorine, magnesium, and manganese. Rice bran is produced when bran layers are removed from brown rice during milling. The nutritional value of rice bran protein is relatively high because of the high lysine content, one of the essential amino acids. Rice bran can be used as a cereal for direct human consumption or as a raw material for production of vegetable oil and host of other applications. However, prior to these applications, it needs to be sterilized and stabilized, since it is perishable and becomes rancid at a fast rate. *Table 6.1* enlists the nutritional composition of rice bran.

Table 6.1: Composition of Rice Bran [*Lu and Luh, 1991*].

Component	Content %
Carbohydrates (a) Hemicellulose (b) Cellulose	 8.7-11.4 9-12.8
Starch	5-15
Oil	15-23
Palmitic Acid	12-18
Oleic Acid	40-50
Minerals, ppm (a) Iron (b) Aluminium (c) Calcium, ppm (d) Chlorine (e) Sodium (f) Potassium (g) Magnesium (h) Manganese (i) Phosphorous (j) Silicon (k) Zinc	 130-530 54-369 250-1310 510-970 180-290 13200-22700 8600-12300 110-880 14800-28700 1700-7600 50-160
Protein %	14-16
Moisture Content, %	6-7

In addition, oryzanol and vitamin E, potent antioxidants, are present in rice bran *[Saunders, 1985]*. The mineral composition of rice bran depends on nutrient availability of the soil in which the crop is grown. *Figure 6.1* describes the process of production of rice bran as a valuable byproduct from rice milling process *[Lu and Luh, 1991]*.

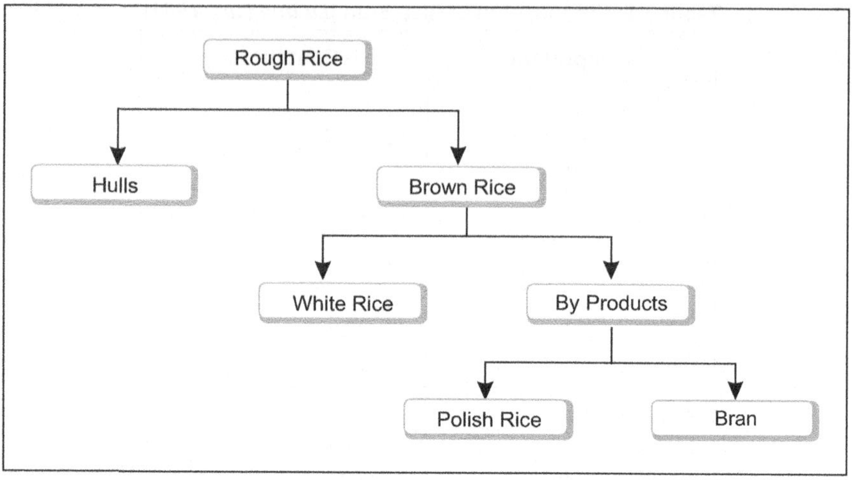

Fig. 6.1: Product fractions during rice milling *[Lu and Luh, 1991]*

i) **Characteristics of Rice Bran:** Rice bran is light in color, sweet in taste, moderately oily, and has a slightly toasted nutty flavor. Texture varies from a fine powder-like consistency to a flake, depending on the stabilization process. In addition to flavor, color, and nutritional properties (protein availability and solubility), other properties, such as, water and fat absorption capacity, emulsifying, and foaming capacity are important factors in the potential use of rice bran in foods. Stabilized rice bran is known as a good source of both soluble and insoluble dietary fiber (25-35%), which is almost twice as much as that of oat bran. The insoluble fiber functions as a bulking agent, while the soluble fiber lowers cholesterol.

The use of rice bran as food and feed is limited, due to its instability caused by hydrolytic and oxidative rancidity. Immediately following the milling process, rapid deterioration of the crude fat in the bran by *lipase* and, to a lesser extent, *oxidase* occurs and makes the bran unfit for human consumption. The naturally occurring *lipase* enzyme in the rice bran hydrolyzes triglycerols (TG), which are primary lipids. The resulting fatty acids increase bran acidity and reduce pH, an off flavor and soapy taste are produced, and functional properties change. Rice bran contains enzyme *lipase* that is site-specific and cleaves the 1,3-site of triglycerols.

Depending on the type of *lipases* present in the bran, storage conditions, and packaging methods, spoilage due to *lipase* continues. Spoilage caused by oxidative rancidity involves a reaction between the lipid and molecular oxygen. The reaction takes place at the double bonds of unsaturated fatty acids and can be accelerated by singlet oxygen, free radicals, metal ions (iron, copper, and cobalt), light, radiation, and enzymes containing a transition metal prosthetic group, such as, *lipoxygenase* (*LOX*). The reactions also depend on the fatty acid composition. Bran, after proper stabilization, can serve as a good source of protein, essential unsaturated fatty acids, calories, and nutrients, such as, tocopherols and ferulic acid derivatives. To process bran into a food-grade product of good keeping quality and high industrial value, all the components causing deterioration must be removed or their activity arrested. Important in this respect is that inactivation of *lipase* and *LOX* enzymes must be complete and irreversible. At the same time, the valuable nutrients must be preserved. *Figure 6.2* explains the spoilage reactions of rice bran *[Lu and Luh, 1991]*.

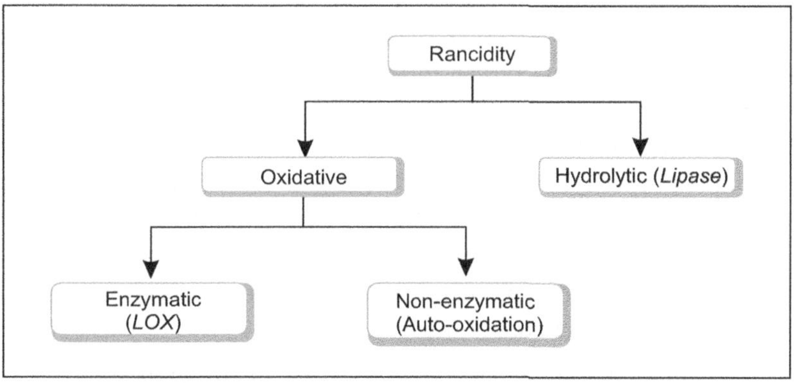

Fig. 6.2: Spoilage reactions in rice bran *[Lu and Luh, 1991]*.

ii) Conventional Methods of Stabilization: Rice bran is a by-product obtained during rice milling. This bran is high in nutrition and fiber. But due to the high activity of the enzyme *lipase,* the bran becomes rancid rapidly. Different thermal methods are used for rice bran stabilization (to inhibit *lipase* activity). Most of the processes involve dry or moist heat treatment. Use of chemicals and irradiation has been unsatisfactory or impractical. The drawbacks common in all heat treatment methods are: (i) severe processing conditions capable of damaging valuable components of bran, (ii) substantial moisture removal, and (iii) complete and irreversible inactivation of enzyme not achieved. It is suggested that moist heat treatment may be more effective than dry heat, but few processes that use steam have achieved satisfactory results. To achieve proper stabilization, every discrete bran particle must have proper moisture content, depending upon the time and temperature of the treatment. Furthermore, moist heat results in agglomeration of bran, resulting in lumpy bran. Extrusion cooking for bran stabilization has been shown to be effective but requires large capital investment. Operating and equipment maintenance costs make the process uneconomical. Use of microwave heating to stabilize rice bran may affect the bran functionalities. Hence, there is a need for an alternative green technology for sterilization and stabilization to obviate the above drawbacks, which seems possible using $scCO_2$.

iii) **Sterilization and Stabilization with $ScCO_2$:** A novel process was developed at IIT Bombay for sterilization and preservation of food products by microbial inactivation and enzyme stabilization using $scCO_2$ at moderate conditions *[Mukhopadhyay and Chakraborty, 2005; 2004]*. In this study a cylindrical 316 stainless steel vessel of 1-liter capacity with rocking facility was used as the steriliser vessel which was washed with soap water, swapped with ethyl alcohol, and dried before conducting the experiments with $scCO_2$ The process developed in this work involves (i) pre-conditioning of CO_2 in the range of 6-30MPa and temperature in the range of 25-60°C, (ii) flushing the preheated sterilizer vessel with CO_2 and charging it with the biomaterial, and pressurising it with preheated CO_2, and (iii) subjecting the pressurised biological material to a processing protocol of pressure cycling at the selected conditions of CO_2 and other process variables. The processed food samples were analysed for microbial count, shelf life, enzyme activity, and other physical characteristics of the final product.

The optimum process parameters have been established based on maximum enhancement of shelf life and maximum enzyme inactivation *[Chakraborty, 2006]*. By optimally controlling the process parameters in the range of 8-10 MPa, temperature 50-60°C, and 3-4 times pressurization and depressurization cycling, the $scCO_2$ processing of rice bran could achieve 99.99% inactivation of microbes and 97% inhibition of *LOX* and *lipase* enzymes with its shelf life enhanced up to 15 days without refrigeration, and without any loss of nutrients. From *Figure 6.3* and *Table 6.2,* it can be seen that the production of free fatty acid owing to the *lipase* activity is lowered significantly indicating that the *lipase* activity has been arrested. Consequently the rancidity of the rice bran is delayed and shelf life enhanced to 15 days without refrigeration, and without any loss of nutrients. The same optimum processing conditions established for sterilization and stabilization of rice bran may be used for another solid food with strong *lipase* as well as *LOX* activities.

Table 6.2: Effect of optimized process parameters on rice bran stability [*Chakraborty, 2006*]

Pressure (MPa)	Temperature (°C)	FFA in Untreated Beans Ao (Absorbance)	FFA in Treated Beans, A (Absorbance)
9	50	1.1621	0.5528
10	50	1.1621	0.4985
10	50	1.2394	0.5418

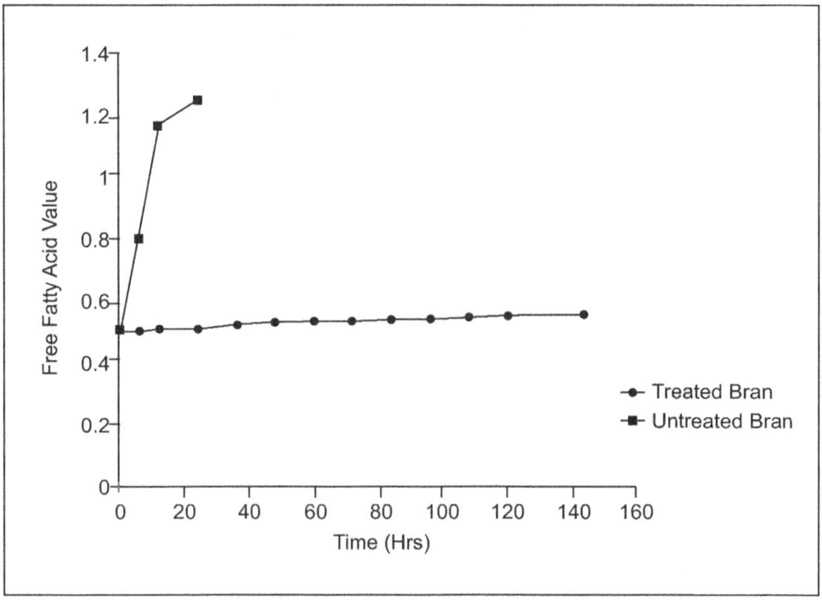

Fig. 6.3: Effect of optimized parameters on FFA in rice bran [*Chakraborty, 2006*]

iv) SCFE of Stabilized Oil from $ScCO_2$-Treated Rice Bran : Rice bran is an oil rich food. The oil extracted from rice bran is very beneficial to health as it lowers cholesterol and prevents acidity. Supercritical fluid extraction (SCFE) was carried out from the $scCO_2$-treated rice bran in order to check the effectiveness of pretreatment of rice bran in improving oil extraction efficiency and the effect of inactivation of *LOX* and *Lipase* enzymes in improving the quality of oil. The extraction was performed at 30MPa, 54°C, with CO_2 flow rate of 0.8kg/h and a batch time of 3h. When no more oil could be extracted from the material, the extraction was stopped and the oil was collected. The yield of the oil extracted by SCFE from the treated rice bran was 9%, which was found to be very high compared to the yield of 4% oil from the untreated rice bran by

Soxhlet extraction from the same starting material. The poor yield of oil by either process is due to the very poor quality of rice bran used for the particular experiment, in view of the fact that good quality rice bran can give an oil yield up to 23%. It is also observed that SCFE of the treated rice bran requires relatively lower pressure of 30MPa of $scCO_2$, as compared to the pressure (40-50MPa) that is required for SCFE from the untreated rice bran for obtaining the same % yield. This is attributed to the morphological changes that take place during the pretreatment of rice bran with $scCO_2$. Moreover the quality of extracted oil is much improved as the oil is sterilized, degummed, enzyme-stabilized, and light in color.

6.1.2 Sterilization and Stabilization of Soybeans

Soybeans are high in protein and a good source of both carbohydrates and fat. They are a rich source of various vitamins, minerals, and beneficial plant compounds, such as, isoflavones. Soybeans are used to prepare innumerable soy products, such as, isolated soy protein concentrates, soy sauce, soy flour, the most important and popular ones being tofu and soy beverages.

i) **Soybeans as the Source for Production of Soymilk:** Soymilk is a plant-based non-dairy beverage, produced from soybeans. It is often consumed as an alternative to milk and is often recommended for lactose-intolerant people. It offers culinary diversity, creamy texture, and a healthful nutritional profile, including flavonoids that exhibit antioxidant, anti-inflammatory, and cardio-protective properties. It is rich in protein and low in saturated fat. Table 6.3 presents a comparison of the nutritional facts of soymilk and cow milk. It is made from soybeans by water extraction of whole soybeans. It is off-white emulsion/suspension containing the water-soluble proteins and carbohydrates and most of the oil present in soybeans. The solubility of soybean proteins in water is strongly affected by pH. Close to 80% of the proteins in raw seeds or unheated meal can be extracted at neutral or alkaline pH, because, as acidity is increased solubility drops rapidly and a minimum is observed at 4.2 to 4.6, which is the isoelectric region of soybean [Berk, 1992].

Table 6.3: Composition and nutritional value of soymilk compared to cow milk *[Berk, 1992]*

Nutritional Value/100g	Cow Milk Whole	Soymilk
Protein, g	3.4	3.6
Fat, g	3.5	2.3
Carbohydrates, g	4.6	3.4
kJ	269	204
Kcal	64	49
Cholestrol, mg	10	0
Lactose, g	4.6	0
Fatty acid, composition		
Saturated, %	63.5	14.0
Polyunsaturated, %	3.0	63.5
Monounsaturated, %	33.5	21.6

ii) Production Processes for Soymilk: Soybean seeds are of two types (i) Elongated which are used as vegetable and (ii) spherical which are the industrial varieties. The different processes for dehulling of soybeans include:

(a) Blanched wet dehulling

(b) Unblanched wet dehulling

(c) Toasted dry dehulling

Different Soymilk producing companies have developed their own production processes apart from the traditional one. The various production processes are:

(a) Traditional Process (as described in *Figure 6.5*)

(b) Soy Technology Systems Process (STS)

(c) INTSOY (Illionois) Process

(d) BUHLER Process for soy micro powder

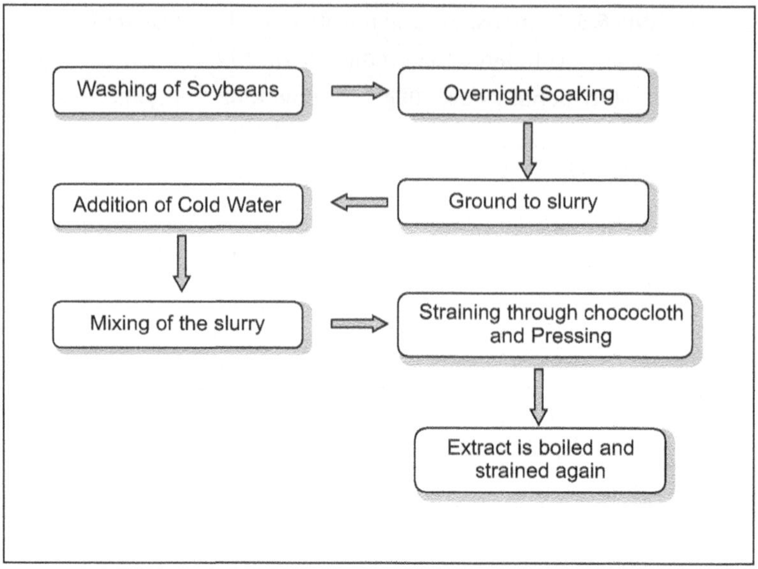

Fig. 6.4: Traditional process for preparing soymilk

The production processes and storage of soymilk influence its quality. During processing the following conditions are required to be followed and a few parameters are needed to be controlled, namely:

i) Cleaning and dehulling without damaging the soybeans

ii) Inactivation of *lipoxygenase* enzyme which causes off-flavor development

iii) Elimination of flatulence causing oligosaccharides

iv) Inactivation of trypsin inhibitors which are present in raw soybeans

v) Maintenance of high protein efficiency ratio

vi) Removal of undesirable smell

vii) Removal of sedimentable solids

Soymilks and related products can be formulated rather freely and may have widely differing compositions which require additional processing. The bulk soybean proteins are globulins *[Berk, 1992]* and are classified into four major fractions as explained in *Figure 6.5.*

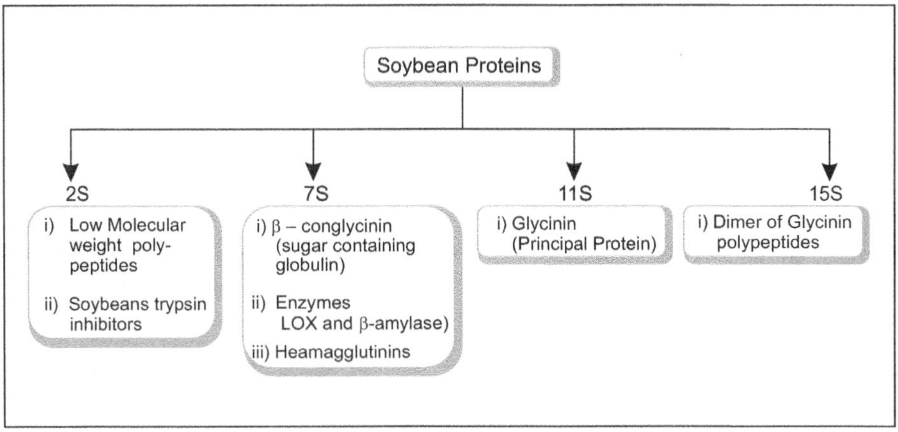

Fig. 6.5: Fractions of soybean proteins. (here S stands for Svedberg units; the numerical coefficient is the characteristic sedimentation constant in water at 20 °C)

During the heat treatment of soymilk, the occurrence of non-enzymatic browning is attributed mainly due to the Maillard's reaction. The measurement of this browning is useful in evaluating the quality of soymilk due to the fact that proteins participate in the reaction. It can thus be used as a quality index for estimation of protein damage in soymilk *[Kwok et al., 1999]*. The slower browning rate in soymilk can be explained by its very low reducing sugar content (glucose) compared to the relatively high lactose content in cow milk, which favors sugar-amine reaction. Another factor which could possibly affect the kinetics of the Maillard's reaction is the structure and conformation of protein molecules which participate in the reaction. The stability of soy proteins can be explained by the complex quaternary structure of the 7S and 11S globular proteins in soymilk as seen from *Figure 6.5*. It is known that these fractions are made up of sub-units, which undergo association and dissociation reversibly. These changes are affected by factors, such as, temperature, pH, and ionic strength. Heating causes further dissociation of both the fractions into sub-units. Further heating causes interaction between the dissociated sub-units to form a soluble polymerized aggregate and insoluble precipitate. Since soy proteins are compactly folded and a high intensity of heat treatment is necessary to unfold the protein molecules completely, it is possible that the kinetics of browning is affected by the kinetics of protein unfolding and dissociation reactions. The fact that soy proteins are relatively heat stable permits the use of a more intense heat treatment in soymilk processing which is essential in order to destroy the anti-nutritional factor (trypsin inhibitor) *[Iwuoha and Umunnakwe, 1997].*

iii) **Deleterious Effect of Enzyme Activity** : The principal inconvenience of traditional soymilk is its beany flavor. This objectionable flavor comes from some ketones and aldehydes, particularly hexanals and heptanals, produced through *lipoxygenase* (*LOX*) catalyzed oxidation of soybean oil. The most crucial enzyme in soybean is *LOX* that catalyzes the oxidation of polyunsaturated fatty acids by molecular oxygen leading to development of rancidity and beany flavor. Soymilk loses its appealing quality if not consumed shortly after production. Attempts to preserve soymilk by conventional methods have remained unsuccessful. Accordingly soybeans are needed to be pretreated before the water extraction process for production of soymilk in order to eliminate its beany flavor.

iv) **Conventional Processes of Pretreatment of Soybeans:** Several approaches have been used to overcome the problem of off-flavors and color in soymilk, which include:

- Heat inactivation of *LOX* in the whole dry bean or during the wet grinding process.
- Need to start with defatted materials (defatted soy flour, soybean protein concentrates, or even isolated soybean proteins)
- Removing the flavor compounds by evaporation for deodorization after they have been formed in soymilk
- Masking the bitterness and off-flavors by sweetening and flavoring
- Developing variety of soybean devoid of *LOX* activity through genetic engineering.

The different pretreatments of soybeans have varying effects on its quality, since proteins are susceptible to very high temperatures, they might undergo varying degrees of denaturation. Studies reveal that soymilk samples stored at ambient temperatures (29 ± 1°C) showed a reduction in protein and fat, and an increase in moisture content and carbohydrate. The relatively pronounced changes in soymilk stored at ambient temperatures are indicative that, at that temperature, biological and chemical changes are encouraged or stimulated to continue, resulting in further degradation of components, in addition to the effect due to the basic processing *[Berk, 1992].*

v) **Sterilization and Stabilization with ScCO_2:** In our first set of experiments it was observed that soymilk could not be directly sterilized and stabilized for shelf life enhancement, by using only scCO_2 for its preservation following our earlier developed process *[Mukhopadhyay* *and Chakraborty, 2004]*. The same process was then modified for microbial inactivation and enzyme stabilization to inactivate the enzymes *lipoxygenase* and *lipase* in solid soybeans prior to dehulling the beans. This includes premixing the beans with a small fraction of water in the ratio of 3:1 and keeping it soaked for an hour. Experiments with $scCO_2$ were designed for pre-treatment of pre-soaked soybeans using the same cylindrical 316 stainless steel vessel of 1-liter capacity as the sterilizer vessel with rocking facility, as mentioned earlier. The objective of these experiments was to pretreat soybeans to optimize the process conditions of $scCO_{2,}$ such that these treated beans have enhanced shelf life and the enzymes are inactivated. The pressures selected for these experiments are 8 -10MPa. The untreated beans had a high enzyme activity. In our experiments the enzyme activity was found to reduce substantially by treating them with $scCO_2$ both at 8MPa and at 10MPa. The *LOX* enzyme activity could be arrested up to 98.6% inhibition at conditions of 10 MPa and 52°C with 4 pressure cycles, as can be seen from *Table 6.5.* It will be seen in Chapter 8 that $scCO_2$ changes the pH of the water around the beans due to formation of carbonic acid. It also diffuses into the beans and swells the beans. The process of pressurization and depressurization cycling renders morphological changes in the treated soybeans facilitating simultaneous tenderizing and dehulling during the $scCO_2$ processing. Post processing these sterilized and stabilized beans can be used readily for preparing soymilk without having to soak them overnight and dehulling, as is practised in the traditional method *[Chatterjee and Mukhopadhyay, 2006].*

Table 6.5: Effect of pressure on LOX activity in soybeans [Chakraborty, 2006]

Pressure (MPa)	Temperature (°C)	Process Intensity	LOX Activity in Untreated Beans, Ao (Absorbance)	LOX Activity in Treated Beans, A (Absorbance)	% Inactivation [(Ao-A)/Ao]*100
8	52	4	0.5886	-0.1789	69.6
10	52	4	0.5886	0.0082	98.6

Figure 6.7 depicts the reduction of *LOX* activity of soymilk prepared from the pre-treated beans from the color loss that was measured in terms of absorbance of dye substrate using a UV- Spectrophotometer at 660 nm over a period of time (as per the procedure explained in Chapter 3). Higher is the *LOX*

activity, higher is the rate of color loss. For the soymilk sample prepared from the soybeans pre-treated at 10 MPa, the color degradation and its rate are the least whereas they are the highest (indicated as –ve) for the untreated sample. There is thus clear evidence of *LOX* inactivation in the $scCO_2$-treated beans, from the rate of change of degradation for the soymilk, as indicated in *Figure 6.7.*

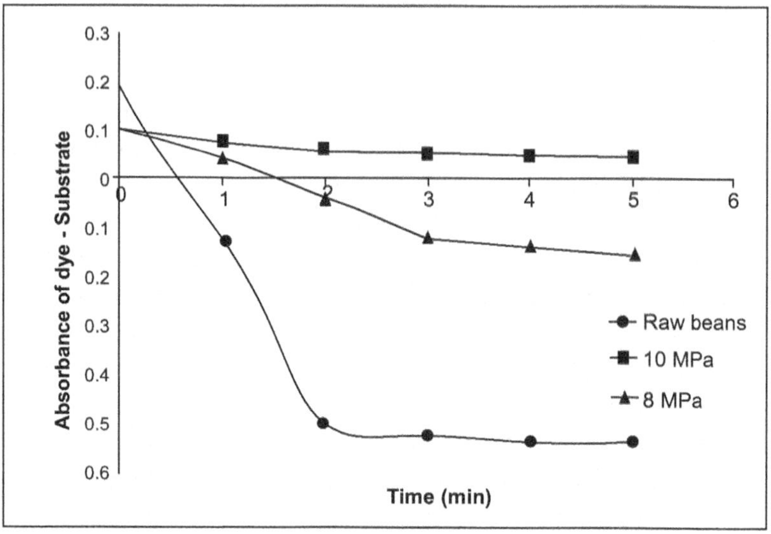

Fig. 6.7: Effect of SC CO_2 pressure at 50°C on Inactivation of *Lipoxygenase [Chakraborty, 2006]*

Further, experiments were carried out with $scCO_2$ at 40, 50, and 60°C for the treatment of soybeans to study the effect of temperature. It was observed that the *LOX* inactivation is highest (99%) with $scCO_2$ at 50°C. Though significant *LOX* inactivation occurs at 40 and 60°C, the *LOX* activity is reduced to a negligible residual activity with treatment with $scCO_2$ at 50°C. This can be attributed to the enhanced mass transfer of the $scCO_2$ fluid at this temperature as compared to higher or lower temperatures due to higher diffusivity and higher solvency respectively. It was thus observed that with some modification in the process conditions, sterilization and stabilization of soybeans could be achieved with high efficiency. Processing with $scCO_2$ makes soybeans devoid of off-flavors and after-mouth feel, making them more acceptable for consumption. Also due to inhibition of trypsin inhibitor activity these beans are safe for consumption by children. Moreover, the treated beans can be used as ready-to-use stock for making soy products. The soy milk prepared from the treated soybeans are devoid of off-taste and chalky mouth feel, and has a higher shelf life than the conventionally prepared soymilk *[Chatterjee and Mukhopadhyay, 2006]* .

6.2 Sterilization of Wheat

Wheat is cultivated worldwide as an important food resource. Wheat grains can be ground into flour and semolina which is the basic ingredient of breads, pastas, and bakery products, making it an important food grain. Thus, it is a staple food to most of the world population, and is usually stored in large quantities post-harvest. The stored wheat grains are susceptible to contamination and spoilage due to various fungi that are present under general storage conditions, which lead to serious deterioration in quality and quantity of the grains, amounting to huge economic loss and loss of important resource. Also, some of the fungi which are known to contaminate wheat, produce mycotoxins as secondary metabolites which are known to be lethal to humans as well as animals. Mycotoxins, such as, ochratoxin and patulin produced by *Penicillium spp.* have been found to be carcinogenic to humans. It is thus imperative that a process of disinfection or sterilization for grains before storage put in place, which will prevent loss of essential resources.

In order to sterilize the wheat grains various technologies, such as, gamma radiation, ozone treatment, pulsed electric field, etc, have been used. However, these methods lead to loss of nutrients, environmental pollution, low germination ratio, etc.

It has been established that $scCO_2$ can eliminate bacterial cells and spores *[Mukhopadhyay and Chakraborty, 2004; 2005]* and microbial biofilms *[Mitchell et al., 2008]*. The study undertaken by *Park et al.* [*2012*], investigated the effects of $scCO_2$ on *Penicillium oxalicum* spores present in wheat grains. Further, they also conducted an analysis to check the germination capability of these sterilized grains. It was noted in the study that an increase in temperature and amount of cosolvent (e.g., water), were significantly correlated to an increase in the inactivation of spores, at a fixed treatment time (30 min in this case). The increase in diffusivity of $scCO_2$ and higher penetration inside cells, under these conditions were responsible for the lethal effect of $scCO_2$ on the fungal cells.

The study also explored, inactivation of *P. oxalicum* spores in suspension and dry spores at 200 bar, 35°C, and a treatment time of 2 h. Their results indicated a 1 log reduction of CFU for the dry spores, whereas a 7-log reduction of CFU for spores in suspension. Thus, it is clearly indicative from this study that water, when added as a cosolvent in $scCO_2$, played a pivotal role in inactivation of *P. oxalicum* spores inoculated on wheat grains.

The study thus concluded that the decontamination method using scCO_2 with water as a cosolvent may be suitable for grains to be directly consumed. However, for grains that are intended for sprouting, this method of sterilization may be a limitation. Further studies are in progress to optimize the operating conditions such that the germination yield of wheat grains is not affected *[Park et al., 2012]*.

6.3 Sterilization of Coconut

Coconut (*Cocos nucifera* L.) has immense health benefits and is popular as fresh-cut fruit served as a snack. The production operations, such as, cutting and washing fasten the microbial growth and accelerate enzyme actions, leading to shortening its shelf life. It is imperative to inhibit microbial and enzymatic activity and to prolong its shelf life without adversely affecting its fresh appearance, original flavor and texture. Several treatments are currently under investigation, such as, sodium chloride treatments, steam blanching, immersion in acid or basic solutions, and use of modified atmosphere packaging.

Ferrentino et al. [2013, 2012] evaluated the effectiveness of processing with scCO_2 as a nonthermal technology for the pasteurization of fresh-cut coconut. It was reported that scCO_2 treatment at 45°C and 12 MPa for 15 min induced a 4 log CFU/g reduction of mesophilic microorganisms, lactic acid bacteria, total coliforms, and yeasts and molds. The hardness of coconut remained unaffected by the treatment but the samples developed an irregular and disorderly microstructure. As per the study carried out, a combination of higher temperature of 45° C and shorter processing time of 15 min was found to be highly effective in inactivating *polyphenol oxidase* enzyme (*PPO*) in coconut. However, the activity of *peroxidase* (*POD*) increased and was found to be resistant to treatment conditions. Interestingly, this treatment left no off-flavor and off-color in the sample, although these enzymes are supposed to be directly related to the deterioration of fresh-cut vegetables and fruits. Further, the study analysed the processed product in storage conditions (at 4° C for 4 weeks) and found that there was no enzymes reactivation. Consequently the samples were devoid of off-flavors and off-colors *[Ferrentino et al., 2013]* and *[Ferrentino et al., 2012]*.

6.4 Sterilization of Meats

Like other solid plant-based foods, an important high protein animal-based food, like meat has also been successfully treated with scCO_2 for sterilization and preservation. The ability of scCO_2 to extract and fractionate fats from ground beef converting them to lower melting point components, in addition, to its effectiveness for removal of cholesterol and inactivation of microbes, has made it a lucrative option which is presently being considered for the meat industry. Two decades ago, a pioneering work was carried out to check for the effectiveness of $scCO_2$ at 42.5°C and 31 MPa for sterilization of ground beef *[Sirisee et al., 1998]* by mixing the meat samples with microbial cultures. It was reported that a longer treatment time was required to inactivate both target microbes: *Escherichia coli* and *Staphylococcus aureus* compared to inactivation of the same microorganisms in a liquid phosphate buffer solution, keeping the treatment parameters same. One log cycle reduction of *E.coli* and *S. aureus* in ground beef was achieved post treatment time of 178 min, while it took 1.7 min to achieve the same result in liquid phosphate buffer solutions. The presence of fats and proteins in ground beef may have played a role in acting as a protective layer for the microorganisms, hindering the bactericidal action of $scCO_2$. Thus, it was concluded that a longer time was needed for inactivation of the microbes in ground beef compared to that in phosphate buffer *[Sirisee, et al., 1998]*. It may be because of some additional time was needed to dissolve the fat layer or to penetrate into it so that $scCO_2$ could come in contact with microbes.

The application of $scCO_2$ for pasteurization of raw chicken meat has also been investigated by researchers. It was confirmed that $scCO_2$ technology is a viable technology for decontamination of raw chicken meat. The inactivation kinetics of *E. coli* with $scCO_2$ alone or in combination with herbs, such as, rosemary or coriander was performed at 40°C and at 80 bar and 140 bar. It was again established that the treatment time was a significant contributory parameter for the sterilization. This is because, an increase in the treatment time was found to result in a higher inactivation, while an increment of pressure from 80 to 140 bar did not increase the inactivation. A 3.25 log reduction in mesophilic microorganisms, 4 log in yeasts and molds, and up to 5 log reduction in *E. coli* were reported to happen. Interestingly, using fresh herbs along with $scCO_2$ did not contribute to any additional inactivation. On the other hand, the use of 0.5% v/w pure essential oils along with fresh herbs (e.g. coriander leaves) resulted in an increase in inactivation. However, it was observed that addition of rosemary herb did not lead to any significant effect. The study also reported that the texture and color of the treated meat resembled cooked samples *[González-Alonso et al., 2020]*.

References:

- Berk; Zeki , 1992, *"Technology of production of edible flours and protein products from soybeans"*. Food and Agriculture Organization of the United Nations, Rome, 1992.
- Chakraborty, Anuradha, *Sterilization and Stabilization of Food Products Using Supercritical Carbon Dioxide,* Ph.D. Dissertation, Indian Institute of Technology, Bombay, 2006.
- Chakraborty, Chatterjee, Anuradha, Mukhopadhyay, Mamata, *"A healthier and Tastier Way to Soy milk," Modern Food Processing,* Vol. 2, No. 2, pp.46-50, 2006,
- Ferrentino, Giovanna, Belscak-Cvitanovic, Ana, Komes, Drazenka, and Spilimbergo, Sara, *"Quality Attributes of Fresh-Cut Coconut after Supercritical Carbon Dioxide Pasteurization,"* Journal of Chemistry Volume 2013.
- Ferrentino, Giovanna, Balzan, Sara, Dorigato, Andrea, Pegoretti, Alessandro, Spilimbergo, Sara, *"Effect of supercritical carbon dioxide pasteurization on natural microbiota, texture, and microstructure of fresh-cut coconut,"* Journal of food science. 77. E137-43. (2012).
- González-Alonso, Víctor, Cappelletti, Martina, Bertolini, Francesca Maria, Lomolino, Giovanna, Zambon, Alessandro, and Spilimbergo, Sara, *"Research Note Microbial inactivation of raw chicken meat by supercritical carbon dioxide treatment alone and in combination with fresh culinary herbs,"* Poultry Science, Vol, Issue 1, pp. 536-545, 2020.
- Iwuoha C.I., Umunnakwe K.E., *"Chemical, physical and sensory characteristics of soymilk as affected by processing method, temperature and duration of storage"*, Food Chem. Jul; 59(3) pp.373-379. 1997, doi:10.10.16/s0308-8146(96)00250-6. AGR:IND21638725.
- Kwok KC, MacDougall DB, Niranjan K *"Reaction kinetics of heat-induced color changes in soymilk"*, J Food Eng.;40:15–20, 1999, doi: 10.1016/S0260-8774(99)00031-X
- Lu, S., and Luh, B. S. *"Properties of Rice Caryopsis"* in B.S. Luh (Eds.), *"Rice production"* (Vol. 2, pp. 389-419). (1991). New York: Van Nostrand Reinhold. http://dx.doi.org/10.1007/978-1-4899-3754-4_11.
- Mitchell et al., *"Resilience of planktonic and biofilm cultures to supercritical CO_2,"* J. of Supercritical Fluids 47 318–325, 2008.
- Mukhopadhyay, Mamata and Chakraborty, Anuradha *"Process for Sterilization of Biomaterials Using Supercritical Fluids"*, Indian Patent no.211305, 2007, Application no. 543/MUM/2004.
- Mukhopadhyay, Mamata and Chakraborty, Anuradha, *"Lipoxygenase-Inactivated and Sterilised Legumes and Cereal Products"* Indian Patent no.282732, 2017, Application no. 540 /MUM/2005.
- Mukhopadhyay, Mamata, Chakrabarty, Anuradha , Invited Poster presentation on *"Novel Processes for Preservation of Fruit Juices, Vegetables and Food products Using Supercritical Carbon Dioxide"*, in the National

Consultative Meet on Fruit & Vegetable Processing, 7 Dec. 2005, NABARD, Mumbai.

- Mukhopadhyay, Mamata, Chakrabarty, Anuradha, *"Sterilization and Stabilization of Food Products with Supercritical Carbon Dioxide at Moderate Conditions"*, in the Proc. of the 11th Intl. Symp. on SFE, SFC and SFP, Pittsburgh, August 1-4, 2004.
- Park, Hyong Seok , Lee, Yong Ho , Kim, Wook, Choi, Hee Jung, Kim, Kyoung Heon, *"Disinfection of wheat grains contaminated with Penicillium oxalicum spores by a supercritical carbon dioxide-water cosolvent system"*, International Journal of Food Microbiology 156 (2012) 239–244.
- Sirisee, U., F. Hsieh, and H. E. Huff, *"Microbial safety of supercritical carbon dioxide processes.,"* J. Food Process. Preserv. 22:387–403, 1998.

❒

7

Sterilization and Preservation of Liquid Foods Using Supercritical CO_2

Growth in fruit and vegetable juice market is driven by increasing demand for healthy food from an increasingly health-conscious consumer base. This is a very transitory food product having a limited shelf life of hours even under the best of circumstances. Hence their sterilization and preservation are important for extending shelf life and to prevent spoilage by killing pathogenic microorganisms and stabilizing deleterious enzymatic activity. There are a number of pathways to deterioration and there have been a number of effective preservation methods that have evolved to eliminate spoilage. An essential requirement of food preservation is to maintain its quality and nutritional attributes while preventing spoilage. In general, the fresher the juice, the higher is the quality; so the standard of excellence is often considered as the freshly prepared real juice. This chapter presents sterilization and preservation of heat sensitive liquid foods with a unique non-thermal processing method with supercritical CO_2 ($scCO_2$) to enhance their shelf life without deteriorating the quality and nutritive value. It also investigates the bactericidal effect of $scCO_2$ treatment for various fruits and vegetable puree/juices for inactivating microorganisms and enzymes to ensure preservation of safety and quality of the product.

M. Mukhopadhayay and A. Chatterjee, *Sterilization and Preservation*,
https://doi.org/10.1007/978-3-031-17370-7_7

7.1 Tomato Puree

Tomatoes (*Solanum lycopersicum*) are the major dietary source of the antioxidant lycopene, which has been linked to many health benefits, including reduced risk of heart disease and cancer. They are also a great source of vitamin C, potassium, folate, and vitamin K. Despite botanically being a fruit, it is generally eaten and prepared like a vegetable. However, for easy availability it is often converted to ready-to-use juice or puree. The quality of tomato juice is characterized by its properties, such as, color, consistency, and flavor. Consistency refers to the viscosity of the product and the ability to hold its solid portion in suspension for the shelf life of the product, presenting almost no syneresis, and is related to cloud, which is defined by fine particles that remain suspended indefinitely as a result of the Brownian motion. As a consequence, tomato juice cloud is responsible for some quality attributes of the juice *[Laratta et al., 1995]*. Tomato puree is usually prepared using a protocol as described in *Figure7.1*.

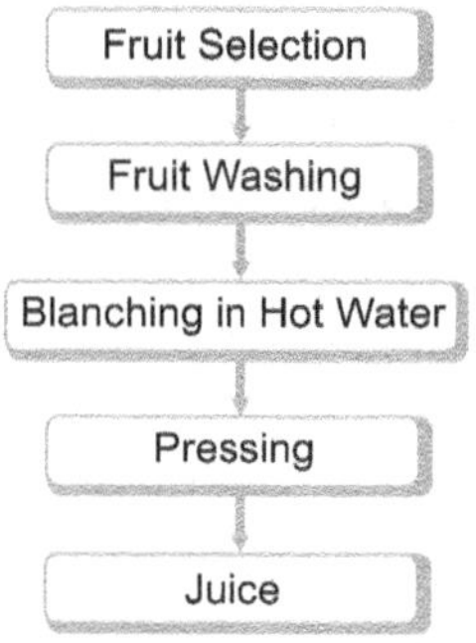

Fig. 7.1: Preparation of tomato juice

7.1.1 Characteristics of Tomato Juice

Freshly prepared tomato juice or puree contains amino acids, up to 49% of which are free glutamic acid. Conventional concentration of tomato juice by heat gives rise to approximate 10-fold increase in free amino acids as a result of protein denaturation and partial hydrolysis. High free glutamic acid levels can have negative effects during storage because of the high reactivity of this acid in the Maillards' type reaction with sugars. The most important component responsible for typical fresh flavor of tomato is n-hexanal. All ultra-high pressure (UHP)-treated samples show a remarkable increase in n-hexanal content, regardless of treatment time and pressure. When present at higher concentrations (than the optimum level of 1-1.2 mg kg^{-1}) this imparts a rancid flavor *[Porretta et al., 1995]*.

Tomato has a particularly complex enzymatic system that is still not completely understood. The enzymes present in tomato are *pectin methyl esterase* (*PME*) or *pectin esterase* (*PE*) and *polygalacturonase* (*PG*). The enzyme *pectin methyl esterase* plays a fundamental role. From a physiological standpoint; *PE* demethoxylates the pectins and is believed to be involved in degradation of pectic cell wall components by *PG* in ripening tomato fruit. Because of discrete thermo resistance of *PE*, it may cause cloud instability when partially inactivated in products, such as, pastes or sauces. In tomato products, cloud very often occurs in the top of a container or appears as a high transmitting serum layer at the center of the container. Rarely does the cloud occur as a very clear serum at the bottom of the container *[De Sio et al., 1995]*.

PE, a cell wall bound enzyme present in tomato fruit, acts on pectin, resulting in cloud destabilization and loss of turbidity in juice. *PE* cleaves the methyl esters of pectin, producing methanol, pectin with a low degree of esterification, and free acid. Once a degree of esterification is reached, divalent cations, such as, calcium can cross-link with these free acid units to other free acids on adjacent pectin molecules. Such cross-linking increases the apparent molecular weights of the aggregates, leading to precipitation of the particles and subsequent clarification of the juice.

The problem of cloud loss has been raised for products, such as, tomato puree (or the pulp obtained by homogenizing the deseeded and deskinned tomato) and tomato juice (or the pulp with added water to make it less viscous), and it is conceivable that early appearance of this effect (from a minimum of 2-90 days) is more rapid than that of fruit juices. This may be due to larger quantity of insoluble material present in this product, to which *PE* is bound *[Laratta et al., 1995]*. Tomato fruit also contains linoleic and linolenic acids as the major constituents in the unsaturated fatty acid pool, and production of volatiles from these can proceed by the sequential actions of *lipoxygenase* (*LOX*) and hydroperoxide *lyase*. *LOX* catalyzes the oxygenation of polyunsaturated fatty acids resulting in a hydroperoxide located at carbon 9 or carbon 13, depending on the isozyme. Hydroperoxide *lyase* causes the cleavage at the carbon containing the hydroperoxide resulting in the formation of an aldehyde and an oxoacid. Hexenal and hexanal are formed from the action of *LOX* and hydroperoxide *lyase* on linolenic or linoleic acid respectively. These are two most significant odor producing fresh tomato volatiles and can be regarded as important constituents in flavor producing pathways *[Riley, et al., 1996]*.

In addition to the fact that *PE* action causes cloud loss of tomato juices, it also prepares a substrate for the enzyme *PG*, which is also present in tomato juice. The enzyme *PG* catalyses the hydrolytic cleavage of the glycosidic α – D

(1 → 4) bonds in the pectin molecule, leading to a decrease in viscosity of the juice due to pectin solubilization *[Fachin et al., 2002]*. *PE* inactivation depends on pH, in particular, the melting temperature. The temperature, where 50% of the enzymatic activity is lost, depends on the pH and increases with it. Thermo resistance of *PE* is higher at higher pH values *[Laratta et al., 1995]*.

7.1.2 Conventional Preservation Processes

Consistency of tomato juice is normally improved using processes, which minimize pectin breakdown by enzymes in pectin and cellulose rich cultivars. The inactivation of *PE* is currently performed by thermal treatment. Thermal processing leads to a series of problems, such as, changes in color, flavor, and vitamin content in processed material. In treatments using high hydrostatic pressure, the total pectin content increases with increasing pressure and is not greatly affected by treatment times. The same occurs with the conventional hot-break treatments for enzyme inactivation, which are normally carried out at decidedly higher temperature (110°C) than those required for pectolytic enzyme inactivation. These methods inactivate enzymes to a lesser extent.

Non-thermal plasma (NTP), has also been reported for the sterilization of tomato juice, in order to retain its fresh flavor and lycopene content. NTP is generated through electric discharge within a gaseous chamber. The reactor may consist of two electrodes and is connected to a high-voltage power supply of 10 kV at 30°C for 5 min. The plasma thus produced is limited within the containment zone [*Ma and Lan, 2015*]. This process seems to be an excellent alternative to classical sterilization methods. However, the shelf life and the sterilization efficiency of the method are yet to be established.

Ultrasound (US) at frequencies in range of 20–100 kHz in the presence of a liquid medium for power transmission has been investigated as an alternative to high temperature- preservation of tomato juice *[Pinheiro et al, 2013]*. However, this technology can cause denaturation of important proteins and polysaccharides present in tomato juice. Treatment by ultrasound is also known to cause protein denaturation *[Vercet et al., 2002]*.

7.1.3 Preservation by Treatment with ScCO2

The principal advantage of using $scCO_2$ for preservation of liquid foods is application of mild conditions of temperature (32-60°C) and pressure (8- 30 MPa). The pressures used in the $scCO_2$ treatment could be about $1/10^{th}$ of the required pressures for high hydrostatic pressure (HHP). This allows retention of quality attributes, nutrients, and other beneficial components, such as, anthocyanins and polyphenols.

Balaban et al. [1994] carried out a systematic research investigation to establish the green technology of $scCO_2$ processing as an alternative technique for preservation of fruit juices and evaluated the efficiency of sterilization. The processing conditions reported were 7.5-34.0 MPa and 35-60°C for a treatment time of 15-180 min. However the process was tested on orange juice, which is a high acid food. The pH of orange juice ranges from 3.3 to 4.2, grape juice has a pH of 3.3, apple juice has an approximate pH value of between 3.35 and 4.0, and tomato juice has a pH of 4.4. It is expected that the $scCO_2$ processing is applicable to all of these juices.

Natural tomato puree treated with $scCO_2$ at lower pressures (7.5-11.0 MPa,) and at lower temperatures (32°-50°C) for up to 24 h, resulted in microbial load reduction by only less than 1 log *[Parton et al., 2007]*. However at higher pressures (20-30 MPa), the $scCO_2$ treatment induced a significant enhancement in shelf life *[Zhao et al., 2018]*. Flavor profile and apparent viscosity were closer to the control (untreated) when treated with $scCO_2$ at 30 MPa and 55° C as compared to the heat-treated sample of tomato juice at 95°C for 20min. Consumer acceptance test indicated that $scCO_2$ juices were preferred to the heat-treated ones, in terms of flavor, taste, and overall acceptability. The $scCO_2$ treatments at 30MPa at 55 °C for 40 and 60 min had positive effects on the content of phenolic compounds, including myricinic acid, ferulic acid, naringin, and chlorogenic acid. A significant increase in the relative content of cis-isomers was observed after the $scCO_2$ treatment at 20 MPa and 55° C. The efficiency of pasteurization process with $scCO_2$ could be increased at higher temperatures. It was reported that 95% of yeast and mold were inactivated after 90 min and 40 min, respectively at 33°C and 55°C at a pressure of 30 MPa. The process with CO_2 is more temperature-dependent rather than pressure-dependent *[Bizzotto et al., 2014]*. Another study recently reported pasteurization of tomato puree with $scCO_2$ at 55°C and 100 bar for 90min, which resulted in 6-7 log reduction. Alternatively, temperature of pasteurization could be lowered to 33°C using nitrogen protoxide (N_2O), also known as, hyponitrous oxide (or laughing gas) along with $scCO_2$ with a treatment time of 60min *[Bizzotto et al., 2021]*. However N_2O is considered to be a health hazard and is normally avoided.

A detailed study was carried out on preservation and stabilization of tomato puree at IIT Bombay using a 1-liter autoclave with rocking facility *[Chakraborty, 2006]* as described in Chapter 6. A number of experiments were carried out in order to analyze the effects of pressure (8-15 MPa), temperature (25-60°C), and, treatment time (1-3 h), and pressurization-depressurization cycling, termed as process intensity (3-5 times) on the sterilization efficiency, cloud stability, and shelf life of tomato puree, and to ascertain the optimum condition for the maximum enhancement of shelf life. It was noted that temperature above 60°C causes color changes in the treated tomato puree. Hence for treatment of tomato puree, the temperature range selected was 25-60°C. Moreover lycopene, one of the important active ingredients for which tomatoes are consumed, is a heat-labile component. Temperatures above 60°C will result in undesirable loss of the valuable antioxidant.

It was also observed that there was very little inactivation with dense CO_2 at 25°C and 8 MPa. This is because at this temperature CO_2 exists as a compressed liquid. As a result, its diffusion into the system and the microbial cells is less. Owing to this low diffusion of CO_2, the sterilizing efficiency achieved by the process is very low. It is found that scCO_2 treatment gives best results at 50°C, and the inactivation efficiency is significantly improved. The inactivation efficiency was found poor even with subcritical gaseous CO_2 at lower pressures of 2 and 3 MPa.

Further, experiments were conducted over a higher range of pressure. The initial experiments were conducted at 30MPa to explore the process performance at a high pressure. The experiments at 30 MPa and 50°C for 60min, did inactivate microbes; however, there was no significant effect on the shelf life. Similar results were observed at 20 and 10 MPa, and higher treatment time of up to 3h. Interestingly, the samples treated at 30 MPa had a tendency to separate into a lower solid puree layer and an upper translucent layer. The experiments were thus modified further, narrowing the pressure to 10MPa at a temperature of 50°C. Also instead of having a long hold time, the experiments were conducted with a hold time of 5min as soon as the system reached the set pressure, followed by immediate depressurization. It was remarkable to find that this pressurization and depressurization sequencing of 4 such pressure cycles enabled bringing down the overall treatment time, as well as pressure and temperatures to 10MPa and 52°C respectively. At the same time, it was possible to achieve near complete microbial inactivation and enzyme stabilization of tomato puree. It was thus established for the first time that it was the use of pressure cycles which helped to lower the requirement of pressure and temperature of treatment with scCO_2 and enabled to reach enhanced sterilization efficiency and an extended shelf life without refrigeration *[Mukhopadhyay and Chakraborty, 2004]*.

The color value of the tomato puree as obtained by the Hunter's Color Lab instrument is presented in *Table 7.1*. The values indicate the color in terms of lightness or brightness (L), redness (a) or greenness (-a) and yellowness (b) or blueness (-b). It is clear that the color of the scCO_2-treated puree is very close to that of the fresh untreated puree. The positive values of a and b further indicate that there is no negative affect of $scCO_2$ processing on the color of the puree at the optimum operating conditions. Also, the observations on the microbial count and other characteristics of the treated tomato puree on the 37th day and 90th day *(Figures 7.2C and 7.2D)* of storage without refrigeration were similar to that of the fresh treated puree and also there was no separation of layers in the puree. This indicates that the treated puree has a stable cloud. The puree usually separates into a watery layer and solid layer due to the action of the enzyme *pectin esterase* (*PE*) which causes cloud instability. It was thus demonstrated that $scCO_2$ processing imparts stabilization of enzyme *PE* as well, in addition to sterilization.

Table 7.1: Effect of $scCO_2$ processing on cloud stability and color of Tomato Puree *[Chakraborty 2006].*

Sample	L value	a value	b value
Fresh untreated puree	27.18	22.22	12.64
Fresh treated Puree	26.95	25.03	13.30
Treated Puree (Day 90th)	26.92	25.05	13.31

Fig. 7.2: Effect of $scCO_2$ treatment of tomato puree: **(**A) and (B) Untreated tomato puree kept at room temperature showing fungal growth on 2nd and 5th day respectively, (C) and (D) $scCO_2$- treated tomato puree kept at room temperature without any growth on 37th and 90th Day *[Chakraborty, 2006].*

7.2 Sugarcane Juice

Sugarcane (*Saccharum officinarum)* contains about 70% of water in which sucrose and other substances are held in solution, whereas the stem contains about 88% by weight of juice in the stem. The remaining 12% represent the insoluble cane fiber component, cane wax, cane fiber, and starch; and being insoluble in the juice, these are removed along with those substances precipitated by lime. Cane wax, $C_{24}H_{50}O$, occurs as a white deposit on the exterior surface of the cane stem, close to each node. Sugarcane juice is obtained using the protocol described in *Figure 7.3.*

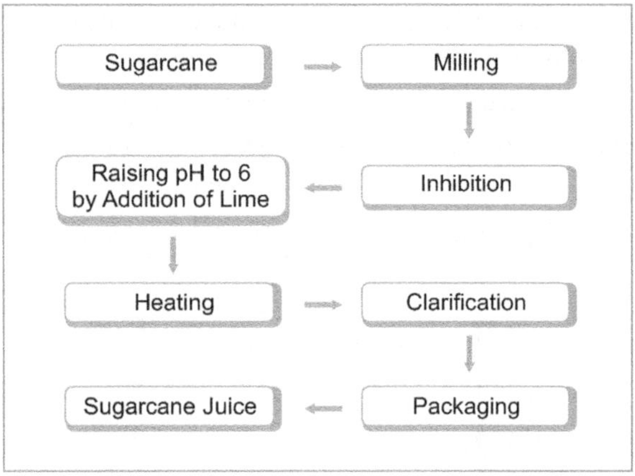

Fig. 7.3: Preparation of sugarcane juice.

7.2.1 Characteristics of Sugarcane Juice

The juice expressed from cane is an opaque liquid covered with froth due to air bubbles entangled in it. Its color varies from light grey to dark green, depending on the coloring matter in the rind of the cane crushed. It contains all the soluble substances like sucrose, fine particles of bagasse, wax, clay (adhering to the cane), coloring matter, and albumen. The proportion of albumen increases when unripe cane or green cane tops are crushed with the ripe cane. The cane juice has an acidic pH between 4.9 and 5.5, which correspond to about 0.2% acidity. Mineral salts in the juice are all derived from the soil and constitute the incombustible ash of the plant. The total quantity of these salts does not exceed 1% by weight of the entire cane plant, proving that the sugarcane removes less mineral matter from the soil than many other crops.

Another major component of the juice is the suspended particles, which form the colloids. The colloids are particles existing in a permanent state of fine dispersion due to which it imparts turbidity to the juice. These colloids do not settle ordinarily unless conditions are altered. The juice is viscous owing to the presence of colloids as waxes, proteins, pentosans, gums, starch, and silica. The gums belong to the polysaccharide group of carbohydrates, and may be divided into two classes: (a) natural gums or pentosans and (b) mucilage and pectic compounds (*www.sugarindia.com*).

Tables 7.2 and *7.3* give the composition of the juice.

Table 7.2: Composition of sugarcane juice (*www.sugarindia.com*)

Parameter	Value (%)
Water	70 to 75%
Sucrose	11 to 16% (avg. = 3.0%)
Reducing sugars	0.4 to 2%
Organic non-sugars	0.5 to 1%
Mineral Matters	0.5 to 1%
Fiber	10 to 16%

Table 7.3: Composition of non-sugars in sugarcane juice (*www.sugarindia.com*)

Acids	Nitrogen Compounds	Coloring matters	Other Organic Non-sugars	Mineral Matters
Organic Acids Glycolic Malic Oxalic Succinic Tannic	Organic Compounds Albumin Albumoses Amines Amino acids Peptones Xanthese Compounds	Soluble Anthocyanin Saccharetin	Soluble pectin gum (xylan)	Mostly soluble alumina, lime, magnesia, potash, soda, sulfur, chlorine
Inorganic acids phosphoric sulphuric	Inorganic compounds of ammonia and nitrogen	Insoluble Chlorophyll	Insoluble cane fibre, cane vax	Insoluble silicates

Browning and changes in flavor and viscosity have been identified as major problems in sugarcane juice during storage. These have resulted to limitations in the distribution of product to bigger and distant markets. Investigation of the factors causing the browning reaction and spoilage reveals that browning occurs almost immediately after juice extraction, mainly due to the enzyme *polyphenoloxidase* (*PPO*). Browning may also be non-enzymatic as it can be brought about by heat treatment of juice. The change in juice viscosity was mainly due to microbial action.

7.2.2 Conventional Preservation Methods

The browning problems are resolved by heat-treating the cane stems prior to juice extraction as well as using anti-browning agents to preserve the color of juice during storage. Heat treatment of the extracted juice is avoided since this causes browning.

Steam blanching of canes prior to juice extraction can preserve the color of the juice as well as prevent microbial spoilage. Blanching not only inactivates the enzyme *PPO* but also kills the microorganisms responsible for spoilage. There is no significant effect of using additives to preserve the color of juice extracted from heat-treated stems. Using the blanching technique, storage can be enhanced from four days to twelve days at 5°C. However, stem blanching is a tedious process and consumes substantial energy. Also, the juice is highly perishable (as the juice spoils within 4-6 hours after crushing if kept at room temperature) due to its high sucrose content, which is ideal for growth of yeasts leading to fermentation.

7.2.3 Preservation and Stabilization with $ScCO_2$

An in-depth study was conducted to investigate the effect of $scCO_2$ treatment on sugarcane juice in the pressure range of 10-16 MPa and temperature in the range of 35-60°C, using the experimental set-up that has been explained in the previous section. A number of experiments were carried out *[Chakraborty, 2006]* to analyze the effects of pressure, temperature, treatment time, and pressurization-depressurization cycle (termed as process intensity). The aim of the study was to determine if sterilization of sugarcane juice could be achieved and also to ascertain the optimum condition for the enhancement of shelf life, since sugarcane juice is known to be highly perishable. For temperatures above 60°C, it was noted that a deep brown color developed and imparted a charred taste, which is undesirable. This change in color is due to charring of sucrose or it is possible that at this temperature the Maillard's browning reaction accelerated. Therefore, the optimal temperature of $scCO_2$ for sterilization and preservation of sugarcane juice was selected to be in the range of 50-55°C.

Based on the previous experiments conducted on liquid foods as presented earlier, it was established that the process of sterilization and preservation of liquid foods works best at a pressure in the range of 8-10 MPa with 4-5 pressure cycles *[Mukhopadhyay and Chakraborty, 2004]*. Further experiments revealed that 80% inactivation of microbes could be achieved at 8 MPa and a near 98% at 10 MPa in one cycle. Subsequently, the pressure cycling was implemented to observe its effect on microbial inactivation and shelf life. A pressure cycle of 4 resulted in an inactivation efficiency of 99.3% with the $scCO_2$ treatment at 10 MPa and 52°C.

It is thus remarkable to find that sugarcane juice could be almost completely sterilized and a shelf life of 17-20 days could be achieved using $scCO_2$ treatment implemented at a pressure of only 10MPa and a temperature of 52°C with 4 cycles of pressurization- depressurization. This is a significant achievement, considering that the ready-to-drink sugarcane juice is difficult to preserve and could be made shelf-stable without refrigeration.

7.3 *Aloe Vera* Juice

Aloe vera (*Aloe barbadensis* Miller) is a medicinal plant that has been used to treat various health conditions. *Aloe vera* is known since ancient times for its remarkable curative properties. While the dried exudates have been used as a cathartic, the juice has been widely accepted since the 4th century B.C. as a traditional medicine for alleviating pain and treating a variety of ailments ranging from burns and lacerations to peptic ulcers, dermatitis, high blood pressure, hair loss and leprosy. *Aloe vera* is a perennial plant with turgid green leaves joined at the stem in a rosette pattern. Aloe leaves are formed by a thick epidermis (skin) covered with cuticles surrounding the mesophyll, which can be differentiated into chlorenchyma cells and thinner walled cells forming the parenchyma (fillet). The parenchyma cells contain a transparent mucilaginous jelly, which is referred to as *Aloe vera* gel. The plant contains two separate juice materials, yellow latex (exudate), extracted from the vascular bundles at the junction between the rind and the fillets, and a transparent mucilaginous gel, extruded from the inner pulp. The yellow latex is mainly composed of aloin, aloe-emodin, and phenols. These are used in mild doses for laxative effect in the cure of constipation. The mucilaginous jelly from the parenchymal cells of the plant is the *Aloe vera* gel *[He et al., 2005]*.

The processing of *Aloe vera* juice, derived from the leaf pulp of the plant, has become a big industry worldwide due to its application in the food industry. It has been utilized as a resource of functional food, especially for the preparation of health drinks which contain *Aloe vera* juice and which have no laxative effects. It is also used in other food products, for example, milk, ice cream, confectionery, and so on.

Figure 7.4 Illustrates the conventional processing of aloe juice.

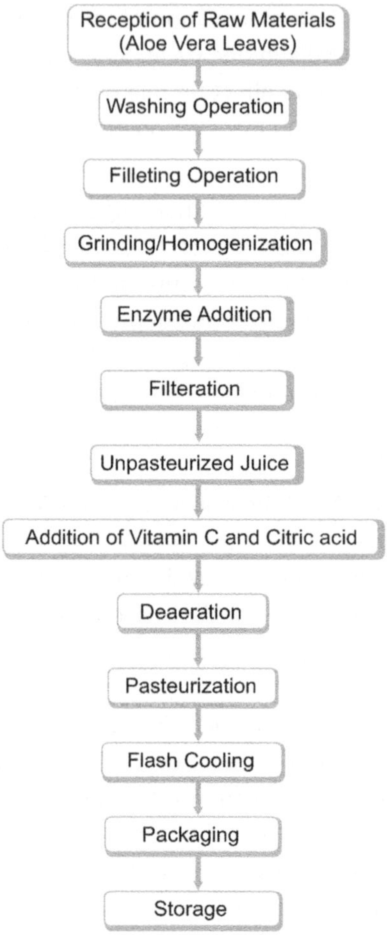

Fig. 7.4: Conventional process for production of single-strength *Aloe vera* gel juice *[He et al., 2005]*

7.3.1 Characteristics of *Aloe Vera* Juice

In recent years, there has been much interest in the biological activity of aloe polysaccharides, which is greater and more diverse than previously realized. Also often mentioned are the antibacterial, antifungal, and even antiviral properties demonstrated by the juice, while its anti-oxidative properties are drawing utmost attention today. The presence of acetyl groups is necessary for activity, because they cover a number of hydrophilic hydroxyl groups and thus enable the molecule to cross-hydrophobic barriers in the cell.

It should also be noted that active glycoproteins have also been demonstrated in aloe juice and may well play some part in therapeutic activity, either immunologically as lectins or as *proteases*, such as, anti-bradykinins *[Reynolds and Dweck, 1999]*.

The chemical composition of Aloe juice is largely dependent on the plant species analyzed. A prominent feature of *Aloe vera* fillet is its high-water content, ranging from 98.5% to 99.5% of fresh matter and the rest as solids, more than 60% of which are polysaccharides. Many studies have reported the presence of polysaccharides as the main component of the fillet with minor amounts of various other components. Acemannan, a mannose-containing polysaccharide, has been reported as the main bioactive substance present in *Aloe vera* fillet. Acemannan, is a lineal polysaccharide composed of (1,4)-linked mannosyl residues, with C2 or C3 acetylated and some side chains of mainly galactose attached to C6 *[Femenia et al., 1999]*.

Acemannan has been incorporated in commercial proprietary wound care products and has been reported to affect wound closure in chronic wounds and aphthous ulcers. Acemannan, if refined further, has been shown to act as an immune-stimulant, displaying adjuvant activity on specific antibody production and enhancing the release of interleukin-1 (IL-1), interleukin-6 (IL-6), tumor necrosis fator-α (TNF-α) and interferon-γ (INF-γ). Release of these cytokines stimulates an increase of up to 300% in the replication of fibroblasts in tissue culture and enhances macrophage phagocytosis *[Ni et al., 2004]*. *Table 7.4* enlists the pharmacological activity of various components of aloe juice.

Table 7.4: Pharmacological activity of aloe components. *[Choi and Chung, 2003]*

Components	Pharmacological Activity
Glycoprotein	Wound Healing Cell Proliferation, Antiallergy
Barbaloin	Purgative
Aloe-emodin, Emodin	Cell Proliferation Anticancer, Anti bacterial, Antioxidant, Genotoxicity, Mutagenicity
Mannose-6-phosphate	Wound Healing, Anti-inflammation
Polysaccharide	Anticancer, Immunomodulation
Acemannan	Immunomodulation Antimicrobial, Antitumor
Aloesin	Cell Proliferation Inhibition of Melanin Synthesis
b-sitosterol	Anti-inflammation, Angiogenesis
Diethylhexylphthalate	Anticancer

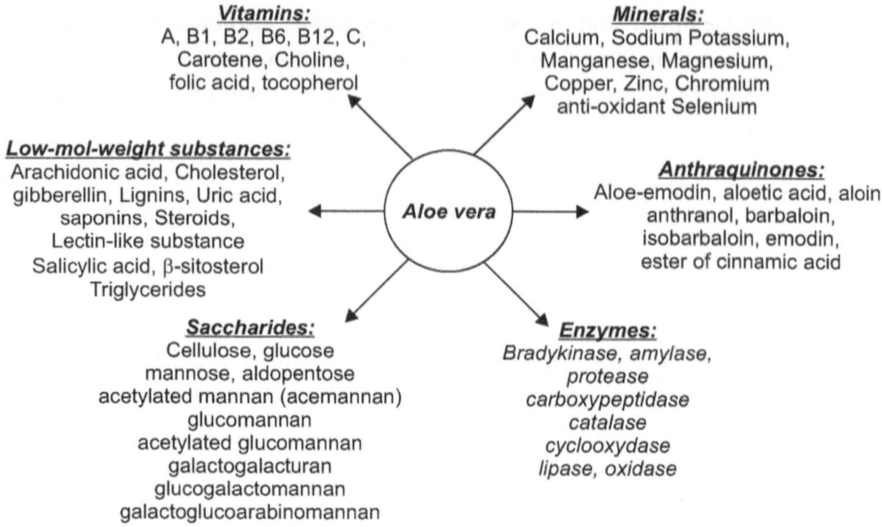

Fig. 7.5: Composition of *Aloe vera* Juice *[Choi and Chung, 2003]*

7.3.2 Preservation of *Aloe Vera* Juice

It is heartening to note that such a high value juice of *aloe* is not stable due to the action of enzyme *cellulase* and microorganisms, and it undergoes degradation within a few hours of harvesting. There are many challenges in the production of a premium grade *Aloe vera* product. The reason for this is the delicate nature of the bioactive component mannose. As soon as harvesting commences, the plant breaks down its own healing substances (polymannose) due to the enzyme *cellulase* found in the tissues of the plant. This natural response enables the plant to repair its wound and to provide a new skin. This process also provides immunity to the leaf and ensures survival of the plant. Without this protective system, the leaf and plant would become susceptible to viruses, bacteria, parasites, and fungus. It is thus absolutely essential to keep the leaves cool (refrigerated) and complete their processing at the earliest, in order to minimize this self-induced degradation. Furthermore, bacteria also pose threat to the polymannose (also known as acemannan) molecules if processing is not effective or delayed. Bacteria not only thrive on the polysaccharide, but also produce enzymes, which further compromise these important complex sugars.

Although color changes have little relation to the therapeutic effectiveness of stabilized juice, they are rarely acceptable psychologically to the user and are totally unacceptable in some products. A simple but efficient processing technique is needed to improve product quality, and to preserve and maintain almost all of the bioactive chemical entities naturally present in the plant during processing *[Ni et al., 2004]*.

For an *Aloe vera* juice, as it starts degrading almost as soon as it is harvested, a proper process for sterilization is necessary in order to prevent the product from fermenting and to guarantee the safety and freshness of an aloe liquid. Methods, such as, pasteurization or High Temperature (over 200 degrees) Short Time (HTST) have been commercially used for the production of *Aloe vera* juices. Such manufacturing practices, though ensure a product of high consumer safety, however, yield low levels of active bio-constituents, as the active ingredient (mannose) is degraded in the processed aloe product.

7.3.3 Preservation and Stabilization of *Aloe Vera* Juice with $ScCO_2$

Aloe juice is very rich in amino acids and has high medicinal and cosmetic values. Its bioactive components are heat labile. Sterilization and preservation of aloe juice require mild treatment conditions and accordingly $scCO_2$ treatment with the process developed earlier *[Mukhopadhyay and Chakraborty, 2004]* was considered most appropriate to conserve the bioactive components of aloe while preserving it. At temperatures above 60-65°C the vitamins and saccharides are prone to denaturation and degradation; hence temperatures below this range were selected for processing *Aloe vera* juice. Moreover, the process parameters, which were earlier evaluated for achieving the maximum sterilization efficiency for bacterial cultures (as elaborated in Chapter 6 and Chapter 8), and were later confirmed to be optimum for preservation of tomato puree and sugarcane juice, were also selected for processing of *aloe* juice (i.e., 10MPa, 52°C and 4 pressure cycles). The $scCO_2$ treatment at 10MPa, 52°C with 4 pressure cycles of each having 5 minutes duration, revealed that the use of pressure cycles enhanced sterilization efficiency to 99.9% microbial inactivation, which led to an extended shelf life *[Chakraborty and Mukhopadhyay, 2005]*.

Figure 7.6 compares the profiles of the microbial count with shelf life of the untreated, the HTST-treated and the $scCO_2$-treated aloe juice samples over a period of time for microbial growth, using the plating method *[Chakraborty, 2006]*.

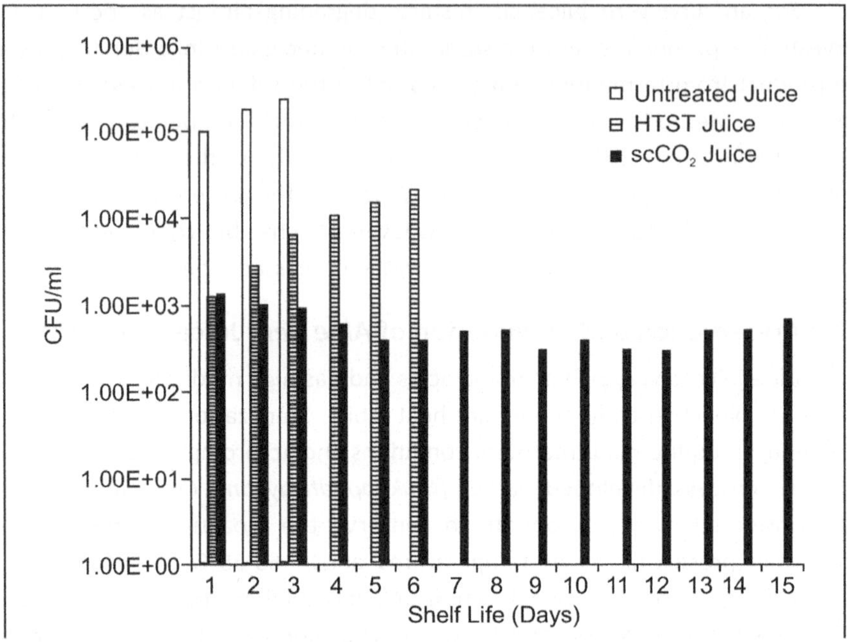

Fig. 7.6: Comparison of microbial count vs. shelf life of fresh untreated juice, commercially available HTST- treated juice, and scCO_2-treated Juice.

It clearly indicates that the microbial growth in the untreated fresh juice and commercially available HTST juice is much faster in comparison to the scCO_2-treated juice (with all three being kept without refrigeration). Obviously, the shelf life of the scCO_2-treated juice is significantly higher than the other two.

It is worth noting that the HNMR analysis on (a) fresh juice and (b) scCO_2-treated juice, as shown in *Figure 7.7* confirms that the components of the treated juice match close to that of the fresh juice. The mannose and polysaccharides are not degraded. This is indicative of the fact that the present process does not negatively affect the aloe constituents. Owing to the mild conditions used for the process the nutritional and medicinal values of the juice are not affected, and the scCO_2 treatment selectively inhibits enzymes and inactivates microorganisms. Consequently, the shelf life of the product is enhanced and the product is shelf stable *[Mukhopadhyay and Chakraborty, 2004]*.

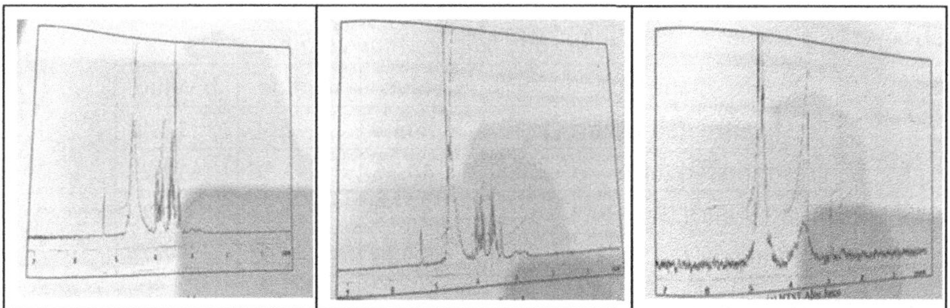

(a) Untreated aloe juice	(b) $scCO_2$ treated aloe juice	(c) HTST treated aloe juice

Fig. 7.7: HNMR:(a) Fresh aloe juice,(b) $scCO_2$-treated juice, (c) HTST treated aloe juice.

Table 7.5 provides the color value of the aloe juice as obtained from the Hunter's Color Lab instrument. The values indicate the color in terms of lightness or brightness (L), redness (a) or greenness (-a) and yellowness (b) or blueness (-b). It can be seen that the color of the treated juice is close to that of the fresh untreated juice color. The -a value close to the original indicates that there is minimal loss in the fresh pale green color of the juice and b values further indicate that there is no negative affect of $scCO_2$ processing on the color of the juice at the optimum operating conditions. Also, the observations on the microbial count and color characteristics of the treated juice on the 30th day of storage without refrigeration are similar to that of the fresh treated juice, and also there is no separation of layers in the juice. This indicates that the $scCO_2$-treated juice has a stable cloud. The untreated fresh juice usually develops a brownish discoloration due to the action of enzymes and microbes, rendering the juice rancid, which did not happen in this case. Also, it can be observed that the color of the HTST- treated juice significantly differs from the fresh juice. This indicates that the HTST-treated juice has undergone some degradation and has lost the fresh appeal.

Moreover, the juice after treatment with $scCO_2$ was less viscous as compared to the untreated fresh juice. This thinning of the juice is however advantageous since enzymes are otherwise added to make the juice less viscous in the commercial processing so that it is drinkable.

Table 7.5: Effect of $scCO_2$ processing on cloud stability and color of *Aloe vera* Juice

Sample	L value	a value	b value
Fresh untreated Juice	43.8	-2.66	12.42
Fresh treated Juice	44.76	-2.36	12.27
Treated Juice (Day 30th)	44.72	-2.33	12.28
Commercial (HTST) Juice	53.89	-1.68	7.79

Thus, it was confirmed that the $scCO_2$ processing of *Aloe vera* juice, by optimally controlling the process parameters at 10 MPa and 52°C with 4 pressure cycles, could achieve 99.9% microbial inactivation, with its shelf life enhanced up to 35 days without refrigeration, without any loss of nutrients and active ingredients like mannose and polysaccharides, and with improved color and consistency *[Mukhopadhyay an Chakraborty, 2004].*

7.4 Cow Milk

Milk and honey are the only articles of diet whose sole function in nature is food. It is therefore not surprising that the nutritional value of milk is high. Milk is a complex fluid with over 100,000 different molecular species found. The constituents of milk may be classified under the following fractions, which are defined as follows:

i) Plasma: That part of milk from which fat has been removed, i.e. Skim Milk.

ii) Serum: Plasma from which the casein micelles have been removed, i.e. Whey.

iii) Solids-Not-Fat (SNF): That part of milk which contains Proteins, Lactose, Minerals, Acids, Vitamins, and Enzymes.

iv) Total Milk Solids: That fraction of milk which combines fat and SNF fraction.

Milk has less water than most of the fruits and vegetables. It can be described as: (i) an oil in water emulsion with the fat globules dispersed in the continuous serum phase, (ii) a colloid suspension of casein micelles, globular proteins and lipoprotein particles, or (iii) a mixture of lactose, soluble proteins, minerals, vitamins and other components. The approximate composition of cow milk is given in *Table 7.6*. The vitamin and mineral compositions of milk are given in *Tables 7.7* and *7.8*.

Table 7.6: Composition of cow milk. *[Dairy Chemistry and Physics, 1999]*

Components	Amount Present (%)
Water	87.3
Milk Fat	3.9
Solid Not Fat	8.8
Protein	3.25 (3/4th casein)
Lactose	4.6
Minerals- Ca, P, Citrate, Mg, K, Na, Zn, Cl, Fe, Cu, Sulphate and Biocarbonate	0.65
Acids- Citrate, Formate, Acetate, Lactate, Oxalate	0.18
Enzymes- Peroxide, *Catalase, Phosphatase, Lipase*	—
Gases - Oxyten, Nitrogen	—
Vitamins- A, C, D, Thiamine, Riboflavin	—

Table 7.7: Vitamin composition in milk. *[Dairy Chemistry and Physics, 1999]*

Vitamin	Content per liter
A (mg, RE)	400
D (IU)	40
E (mg)	1000
K (mg)	50
B1 (mg)	450
B2 (mg)	1750
Niacin (mg)	900
B6 (mg)	500
Pantothenic acid (mg)	3500
Biotin (mg)	35
Floic acid (mg)	55
B12 (mg)	4.5
C (mg)	20

Table 7.8: Mineral content in milk *[Dairy Chemistry and Physics, 1999]*

Mineral	Content per liter
Sodium (mg)	350-900
Potassium (mg)	1100-1700
Chloride (mg)	900-1100
Calcium (mg)	1100-1300
Magnesium (mg)	90-140
Phosphorous (mg)	900-1000
Iron (mg)	300-600
Zinc (mg)	2000-6000
Copper (mg)	100-600
Manganese (mg)	20-50
Iodine (mg)	260
Fluoride (mg)	30-220
Selenium (mg)	5-67
Cobalt (mg)	0.5-1.3
Chromium (mg)	8-13
Molybdenum (mg)	18-120
Nickel (mg)	0-50
Silicon (mg)	750-7000
Vanadium (mg)	310
Tin (mg)	40-500
Arsenic (mg)	20-60

7.4.1 Characteristics of Cow Milk

The fat content of milk is of economic importance because it is sold on the basis of its fat content. Milk fatty acids originate either from microbial activity in the rumen and transported to the secretory cells or from synthesis in the secretory cells. The main milk lipids are called the triglycerides (98.3%), comprising glycerol backbones binding up to three different fatty acids. Milk also contains phospholipids (0.8%) and cholesterol (0.3%). At room temperature the lipids are solids and hence referred to as fats, the melting point of milk fat being 37°C. Crystallization of milk fat largely determines the physical stability of the fat globule and the consistency of high fat dairy products *[Whitaker, 1994]*.

More than 95% of the total milk lipid are in the form of globules ranging from 0.1 to 15µm in diameter. These droplets are covered by a thin fat globule membrane (FGM, 8-10nm in thickness). This decreases the lipid-serum interfacial tension to very low values (1-2.5 mN/m), preventing the globules from immediate flocculation and coalescence, also protecting them from enzymatic action. The Stoke's law predicts that if raw milk is left to stand then the fat globules will cream due to the differences in the densities between fat and plasma phases in the milk.

The primary structure of milk proteins consists of a polypeptide chain of amino acid residues joined together by peptide linkages, which may also be cross-linked by disulfide bridges. The nitrogen concentration of milk is distributed among caseins (76%), whey proteins (18%), and non-protein-nitrogen (6%). The distinguishing property of the caseins is their low solubility at pH 4.6 and their main biological function is to carry highly insoluble CaP to mammalian young in liquid form. The proteins appearing in the supernatant after precipitation at pH 4.6 are collectively called as the whey proteins. They have good gelling and whipping properties.

The action of enzymes on milk is very specific. Milk contains both endogenous and exogenous enzymes. The endogenous enzymes belong to the *hydrolases* group while the exogenous ones are *lipases* and *proteinases* produced by the psychrotrophic microorganisms. These enzymes and microorganisms are not desirable as they shorten the shelf life of milk. However, some enzymes and microorganisms may be useful to convert milk to other milk products e.g: the enzyme *Plasmin* plays a role in developing the flavor in cheese, the enzyme secreted by *Lactobacillus acidophilus* helps to prepare yogurt, and *Streptococci diacetilactis* produces diacetyl which gives the unique flavor to ice creams *[Whitaker, 1994]*.

7.4.2 Preservation and Pasteurization of Milk

Milk marketing standards, e.g., Pasteurized Milk Ordinance (PMO), and standard procedures ensure a wholesome product entering the food chain. Pasteurization of milk is a standard practice to eliminate pathogens in the products sold. The growth of many kinds of bacteria can be retarded or inhibited by refrigeration and freezing. However, refrigeration alone cannot kill all the bacteria. Pyscrophiles are usually nonpathogenic but are able to multiply under cold conditions. These attack the proteins and fats, causing off-flavors and shortening shelf life. The maximum allowable number of bacteria in milk that can be marketed for human consumption is 10^5 CFU/ml *[Holzapfel et al., 1995; Whitaker, 1994]*. Pasteurization is the most common preservation technique used for milk. Applying temperatures below its boiling point inactivates the organisms. For milk it is 63°C for 30min or 72-75°C for 15-20s. Chemical preservatives like nitrates, nitrites, and sulfides are also used as preservatives. These act either as direct poison for the organisms or lower the pH creating an acidic environment to prevent growth. However, a number of health side effects are associated with the use of these preservatives. Ultra high temperature (UHT) used for milk involves heating at 140°C for a few seconds followed by rapid cooling. However, the process destroys the heat labile vitamins like vitamin C *[Gedam et al., 2007]*.

Though many technologies for preservation of milk have evolved, none has been successful in enhancing shelf life of milk without disturbing the nutritional value. Pasteurization certainly inactivates microorganisms and enzymes, but the effect is short-lived. Application of UHT is also undesirable as it denatures proteins and also aggregation of casein micelle occurs at high temperatures. High temperature fails to ensure inactivation of enzymes like *Plasmin* or *Lipases,* which again damage milk. Chemical preservatives like nitrates, nitrites, and sulfites are associated with health hazards.

Pasteurization destroys or retards most of the microbes and endogenous enzymes *(Alkaline phosphatase)* present in milk. Although most psycrophiles are killed by pasteurization, the undesirable effect of their activity remains. This effect results reduced yield of milk products, shortened shelf life, off-flavors, rancidity, etc. This treatment too renders milk a shelf life of only 16 hours without refrigeration and also denatures the heat labile proteins. Hence, the problem of short shelf life persists and this method deteriorates the nutritional value of milk. Thus, there is a need to adopt an alternate technology to preserve milk.

7.4.3 Preservation of Cow Milk by ScCO$_2$ Treatment

A detailed study was carried out to explore the possibility of using the scCO$_2$ technology in place of the high temperature pasteurization for preservation of milk and to evaluate the effect of scCO$_2$ on the stability of processed cow milk. During treatment with scCO$_2$, carbonic acid is formed, which lowers the pH of the food system to 4.0 to 3.3 depending on the pressure of scCO$_2$ and the treatment time. The treatment of soymilk with scCO$_2$ using the process presented earlier for liquid foods caused curdling resulting tofu rather than the sterilized and stabilized soymilk *[Chakraborty, 2006]*. That was the reason why soybean is first treated with scCO$_2$ and then the treated beans are used as the starting material for preparation of sterilized soymilk.

The several experiments were carried out in the same experimental set-up as already described in previous sections, using the same optimal process parameters determined for other liquid foods, namely, 8-10 MPa, 50-60°C and 4-5 pressure cycles *[Mukhopadhyay and Chakraborty, 2004]*. However when milk is subjected to the treatment with the mildest process intensity, the pH of milk is lowered, due to which the proteins denature as the isoelectric point of milk is at pH 5.4. At this point, it is worth noting that as CO_2 diffuses out of the system on depressurization, the proteins renature or recoil. Consequently, there is no separation owing to the reversible nature of protein unfolding with the mildest process intensity *[Chakraborty, 2006]*.

It was found that the inactivation efficiency of the cow milk using $scCO_2$ was much less using one pressure cycle at 8MPa and 50-55°C. This is because, at this condition, the dissolution of $scCO_2$ in the system is not high enough to result high microbial kill. Therefore, the optimal pressure of $scCO_2$ for sterilization and preservation of cow milk is established to be in the range of 8-10 MPa., 52°C and 35 minute treatment time in one cycle only. The milk tasted like a carbonated drink immediately after treatment with $scCO_2$. This carbonated taste gradually subsides as CO_2 diffuses out of the system. Based on the experimental findings, the optimal process parameters for milk were determined but the pressure cycling could not be implemented for milk.

Interestingly, it was observed that the process results in separation of whey and formation of cottage cheese from the treated milk, with more than one pressure cycling, or if stirring is used or dip tube is used for CO_2 inlet. These are valuable shelf-stable milk products of commercial interests. The amount of CO_2 that diffuses into the system in this case, is much higher than that obtained without the pressure cycling. Moreover, the dissolved CO_2 remains in the system for a longer time and does not diffuse out of the system immediately after depressurization. As a result, separation of casein and whey was observed, as the denaturation became irreversible. Similarly, when a dip tube or stirring is used, more CO_2 is dissolved due to enhanced mass transfer resulting higher drop in pH and causing irreversible denaturation of proteins. Therefore, enhancing mass transfer has adverse effect in the case of milk, as the food system undergoes a change from one form to another.

Since high process intensity or any other form of enhancing mass transfer could not be applied in case of milk, the bactericidal and shelf life-enhancing capacity of $scCO_2$ was limited. Oil or fat content in milk provides an extra barrier to the bacterial cells protecting it from the direct contact with environment with dissolved CO_2. Therefore, the shelf life enhancement achieved is limited to 5 days without refrigeration.

7.5 Apple Juice

Apples (*Malus domestica*) are rich in phyto-chemicals, particularly polyphenols and flavonoids which may help protect human cells from disease-promoting oxidative stress and inflammation. Apple juice is high in water content of (about 88%) and tastes good, making it a good choice for hydrating especially for those who are sick and at an increased risk of dehydration. To avoid side effects, it is diluted to half-strength when using it to rehydrate after an illness. Drinking apple juice may reduce the risk of heart disease by protecting LDL (bad) cholesterol from oxidation and prevent cancer by increasing the antioxidant activity in blood. Preliminary human research suggests that it may improve behavior and mental health in Alzheimer's disease. It is no wonder when a proverb says ' An apple a day, keeps doctor away'.

7.5.1 Characteristics

Cloudy apple juice is considered as a more beneficial option to human health as compared to consumption of clear apple juice, due to the former's higher nutrient content. This high nutrient content is lost during various available processing conditions. The main problem with cloudy apple juice during production and storage is the retention of color and the cloud stability. The rapid discoloring of cloudy apple juice is the effect of enzymatic browning, due to the action of *polyphenol oxidase* (*PPO*) which catalyzes oxidation of phenolic compounds. The exact mechanism of cloud stability in juices still needs an in-depth study. It is most likely linked to the *pectin methylesterase* (*PME*) activity. The electrostatic repulsion by negative charges present in the partly demethylated pectin due to presence of galacturonic residues could be the reason behind cloud stability. The activity of *PME* causes pectin demethylation along with the formation of insoluble calcium pectate gels which may result in precipitation causing clarification and the loss of turbidity. *Tables 7.9 and 7.10* present the chemical composition and major constituents of apple juice respectively.

Table 7.9: Chemical composition of apple juice *[Alvarez et al., 2000]*

Compound	Concentration (g/l)
Water	860-900
Sugars	100-120
Fructose	46-70
Sucrose	27
Glucose	20
Malic acid	3-7
Pectin	1-5
Starch	0.5-5
Polyphenols	1
Proteins	0.6
Vitamins (mainly ascorbic acid)	0.05

Table 7.10: Major constituents of apple juice *[Savatović et al., 2009]*

Component	Raw Juice
Soluble solids (g/100g)	15.15
Acidity (g malic acid/100g)	0.21
Reducing sugars (g/100g)	11.41
Total sugars (g/100g)	13.67
Brown Component (mg $K_2 Cr_2 O_7$/ml)	117.5

7.5.2 Conventional Preservation Process

Currently, HTST pasteurization is the most commonly used method for heat treatment of apple juice. For example, apple juice is treated by HTST at a temperature in the range of 77-88°C for a treatment time of 25 - 30 s. As an alternative to the use of heat treatments for pasteurization, the Pulse Electric field (PEF) treatment is recommended as it is classified as a non-thermal food processing method because of very low increase in temperature during application. PEF cannot cause protein coagulation or gelatinization of starch. Moreover, covalent chemical bonds are unaffected, hence the nutrients remain intact. In contrast to the conventional treatment process, PEF can be performed at room temperature in a continuous fashion in a short period of time within seconds. PEF is efficient in microbial inactivation, as well as in preserving some apple juice quality attributes, such as, pH and color. Generally, PEF retains most of the volatile compounds responsible for color and flavor of the apple juice, e.g., acetic acid, hexanal, butyl hexanoate, ethyl acetate, ethyl butyrate, methyl butyrate, and hexyl acetate. In addition, more polyphenols are retained by PEF than HTST. However, in terms of shelf life, it was found that thermal pasteurization (HTST) of apple juice is significantly more efficient in bacterial decrease than PEF. Nevertheless, PEF treated samples had relatively less log count of microbial load than that of the HTST-treated samples, though the apple juice had better flavor and color *[Korma et al.,2016]*.

Ultraviolet-radiation treatment has been successfully used for reducing the microbial load in different fruit juices and nectars, and under optimal treatment conditions. More than 5-log reduction of *E. coli,* in fruit juices could be achieved. The minimum treatment parameter for clear apple juice was 230 J L^{-1} of UV dosage. High pressure homogenization (HPH) is yet another promising nonthermal technology being explored for preservation of fruit juices. HPH acts via a combination of spatial pressure and velocity gradients, turbulence, impingement, cavitation, and viscous shear, leading to the microbial cell disruption. The process, however, requires very high pressures in the range of 250-350MPa to achieve 5 log reduction of microbes in juices *[Rupasinghe and Yu, 2012]*.

7.5.3 Preservation with $ScCO_2$ Treatment

Treatment with $scCO_2$ was employed for sterilization and preservation of 'Golden' delicious apple juice to inactivate microorganisms and certain enzyme *PPO* using a very high pressure of 50MPa at a temperature of 50ºC *[Illera et al., 2017]*. Though the initial pH of the juice of 3.89 immediately underwent a decrease to reach a pH of 3.74 post treatment with $scCO_2$, however, after 2 hours of depressurization, the pH returned to its initial value. The color of the juice also did not visually change. It was found that there were no significant differences in the content of total polyphenol compounds and in the antioxidant capacity in the juice post treatment. It was thus concluded that $scCO_2$ treatment is a valid alternative for a near total inactivation of micro flora and *PPO* in cloudy 'Golden' delicious apple juice. However, the enzyme *PME* was more resistant to inactivation with $scCO_2$ at 50Mpa and 50°C. *[Illera et al., 2017]*. As mentioned in earlier sections, this may be achieved by the application of pressure cycling, along with lowering of treatment pressure for achieving the same results in an efficient way.

The effect of pressure of $scCO_2$ on sterilization and stabilization of apple juice was evaluated in another study at a wide range of pressures, namely, 8, 15, 22, and 30 MPa at a temperature of 55°C for a period of 60min treatment time *[Gui et al., 2006]*. A significant decrease in *PPO* activity and cloud stability could be achieved in the treated samples at the highest pressure of 30 MPa,. It was found that the samples treated at 30 MPa showed a significant reactivation of *PPO* after the first week of storage at 4°C, thereafter, it remained constant. This reactivation of *PPO* was not found to have any deteriorative effect on the juice quality. The color of the $scCO_2$ treated juice at 30 MPa also had acceptable color and did not degrade during the 4 weeks of storage at 4°C. However application of pressure cycles at a lower pressure and temperature of $scCO_2$ may give beneficial results, while also enabling the storage at room temperatures.

Apple juice prepared from 'Annurca' apple puree was treated with $scCO_2$ in a batch system at much lower pressures between 7-16 MPa, and temperature between 35-60°C for 40-60 minutes and compared with the thermal process carried out over a range of temperature of 35 -85°C and treatment time of 1-140 minutes. It was reported that microbial inactivation of 5-log reduction of natural flora in apple juice could be achieved by the $scCO_2$ treatment at 16 MPa and 60°C for 40 minutes, as could be achieved by the thermal process at 85°C for treatment time of 60 minutes. It was observed that temperature played a more significant role in improving the sterilization efficiency, with inactivation significantly enhanced when it increased from 35°C to 60°C than the pressure by increasing it from 7 to 16 MPa. However, both temperature and pressure could be lowered to 35°C and 13 MPa respectively, by incorporating 6 pressure cycles of $scCO_2$ treatment for 10 minutes each, to achieve the same 5-log reduction of natural flora in apple juice *[Ferrentino et al., 2009]*. It is thus concluded that $scCO_2$ treatment could be a promising alternative technology in the beverage industry to produce juices with fresh-like characteristics while extending shelf life and safety considering the huge consumption and popularity of fruit juices.

References

- Alvarez, S., Riera, F., Álvarez, R., Coca, J. Cuperus, S F.,.Th. Bouwer, G. Boswinkel, R. V. Gemert, J. Veldsink, L. Giorno, L. Donato, S. Todisco, E. Drioli, J. Olsson, G. Trägårdh, S. Gaeta, L. Panyor; *"A new integrated membrane process for producing clarified apple juice and apple juice aroma concentrate"*, Journal of Food Engineering, 2000, DOI:10.1016/S0308-8146(00)00139-4.
- Balaban, Murat,O., Pekyardimci, S., Chen, C.S., Arreola, A., Marshall, M.R., *"Enzyme Inactivation by Pressurized Carbon Dioxide"* Developments in Food Engineering pp. 855-857, 1994, In: Yano T., Matsuno R., Nakamura K. (eds) Developments in Food Engineering, Springer, Boston, MA.
- Bizzotto Silvia, Vezzù Keti, Bertucco Alberto, Bertoloni Giulio. Non-Thermal Pasteurization of Tomato Puree by $ScCO_2$ and N_2O. Thesis in Chemical Engineering, Universitadegli Studi di Padova Dipartimento di Principi e impianti di Ingegneria Chimica (2021).

- Bizzotto, Silvia, Vezzù, Keti, Bertucco, Alberto and Bertoloni, Giulio, *"Non-Thermal Pasteurization of Tomato Puree by Supercritical CO_2 and N_2O"*, 2014.
- Chakraborty, Anuradha, *"Sterilization and Stabilization of Food Products Using Supercritical Carbon Dioxide"* Ph.D. Dissertation, IIT Bombay, 2006.
- Chakraborty, Anuradha and Mukhopadhyay, Mamata *"The Juicy Trail Using Supercritical Carbon Dioxide to Preserve Aloe Juice"*, Modern Food Processing, Nov-Dec, 2005, 36-39.
- Choi S, Chung *"A review on the relationship between Aloe vera components and their biological effects"*. Semin Integr Med 1:53–62, MH (2003).
- Dairy Chemistry and Physics, 1999, https://www.uoguelph.ca/foodscience/book/export/html/1883.
- De Sio Francesco, Dipollina Giuseppe, Villari Gerardo, Loiudice Roberto, Laratta Bruna, Castaldo Domenico. *"Thermal resistance of pectin methylesterase in tomato juice"*, Food Chemistry, 52, pp. 135-138. (1995), 10.1016/0308-8146(94)P4192-I.
- Fachin Diana, van Loey Ann, Ly Nguyen Binh, Verlent Isabel, Oey Indrawati, Hendrickx Marc. (2002). *"Comparative Study of the Inactivation Kinetics of Pectinmethylesterase in Tomato Juice and Purified Form"*. Biotechnology Progress. 18. pp. 739-44. 10.1021/bp0155080.
- Femenia A, Sànchez E.S, Simal S, Rossello C., *"Compositional features of polysaccharides from Aloe vera (Aloe barbadensis Miller) plant tissues"*, Carbohydr Polym.; 39:109–17, 1999.
- Ferrentino, Giovanna, Bruno, Mariacarmela, Ferrari, Giovanna, Poletto, Massimo and Balaban, M.O., *"Microbial inactivation and shelf life of apple juice treated with high pressure carbon dioxide"* J Biol Eng. 2009; 3: 3. Developments in Food Engineering pp 855-857.

  Gedan, Kishor, Prasad Rajendra, and Vijay V.K., *"The Study of UHT Processing of Milk : A Versatile option for Rural Sector"*, World Journal of Dairy & Food Sciences 2(2) 149-53 (2007).
- Gui Fenqi, Wu Jihong, Chen Fang, Liao Xiaojun, Hu Xiaosong, Zhang Zhenhua, Wang Zhengfu. *"Change of polyphenol oxidase activity, Color, and browning degree during storage of cloudy apple juice treated by supercritical carbon dioxide"*. European Food Research and Technology. 223. 427-432. (2006), 10.1007/s00217-005-0219-3.
- He Qian, Changhong Liu, Kojo Eshun, Zhang Tian, *"Quality and safety assurance in the processing of Aloe vera gel juice."* Food Control. 16. 95-104, (2005), 10.
- Holzapfel, W.H., Geisen, R, and Schillinger, U., *"Biological preservation of foods with reference to protective cultures, bacteriocins and food-grade enzymes"*, International Journal of Food Microbiology, Vol.24, 343–362 (1995).
- Illera, Alba Ester, Teresa, M, Sagrario Sanz, Ángela Beltrán, Solaesa, G. and Melgos, Rodrigo, *"Enzymatic inactivation of apple juice using HPCD (High Pressure Carbon Dioxide) technology and its effect on the quality*

parameters of the juice," Ponencia presentada en: 16th European Meeting on Supercritical Fluids, 25 a 28 de abril de 2017, Lisboa.

- Korma S.A., Kamal-Alahmad, Ali AH, Shoaib M, Abed SM, Yves H, et al. *"Application of Pulsed Electric Field Technology in Apple Juice Processing",* Austin J Nutri Food Sci. 2016; 4(2): 1080.
- Laratta Bruna, Fasanaro G., De Sio Francesco, Castaldo Domenico, Palmieri A., Giovane Alfonso, Servillo Luigi *"Thermal inactivation of pectin methylesterase in tomato puree: Implications on cloud stability".* Process Biochemistry. 30. 251–259. . (1995). 10.1016/0032-9592(95)85006-6.
- Ma, T. J. and Lan, W. S., *"Effects of non-thermal plasma sterilization on volatile components of tomato juice",* Int. J. Environ. Sci. Technol. (2015) 12:3767–3772.
- Mukhopadhyay Mamata and Chakraborty Anuradha, *"Process for Sterilization of Biomaterials Using Supercritical Fluids",* Indian Patent No., 211305, dated 24.10. 2007, Application No. 543/MUM/2004.
- Mukhopadhyay, M., Chakrabarty, A., Invited Poster presentation on *"Novel Processes for Preservation of Fruit Juices, Vegetables and Food products Using Supercritical Carbon Dioxide",* in the National Consultative Meet on Fruit & Vegetable Processing, 7 Dec. 2005, NABARD, Mumbai.
- Mukhopadhyay, M., Chakraborty, A., *"Sterilization and Stabilization of Food Products with Supercritical Carbon Dioxide at Moderate Conditions",* in the Proc. of the 11th Intl. Symp. on SFE, SFC and SFP, Pittsburgh, August 1-4, 2004.
- Mukhopadhyay, M, Chakraborty, A., *"Preserving Goodness of Aloe Vera Using Supercritical Carbon Dioxide",* at the National Seminar : Botanical Products in New Millenium-Developments and Challenges, 5-7 February, 2005, Jaipur, University of Rajashtan.
- Ni Y., Turner D., Yates K.M., Tizard I. *"Isolation and characterization of structural components of Aloe Vera L. leaf pulp",* Int Immuno-pharmacol; 4:1745–1755, (2004).
- Ondrey and Kamiya, 2000.
- Parton, Tiziana, Bertucco A., Bertoloni Giulio. *"Pasteurisation of grape must and tomato paste by dense-phase CO_2".* Italian Journal of Food Science. 19. 425-437, (2007).
- Pinheiro, Joaquina, Goncalves, Elsa M., and. Silva, Cristina L.M., *"Alternative technologies for tomato post-harvest quality preservation",* 2013, CAB Reviews Perspectives in Agriculture Veterinary Science Nutrition and Natural Resources, 8(61):1-15.
- Porretta, Sebastiano , Birzi, Alessandra, Ghizzoni, Claudio, Vicini. Enzo *"Effects of ultra-high hydrostatic pressure treatments on the quality of tomato juice",* Food Chemistry, Volume 52, Issue 1, Pages 35-41, ,(1995) ISSN 0308-8146, https://doi.org/10.1016/0308-8146(94)P4178-I.

- Reynolds T, Dweck AC Aloe vera leaf gel: a review update. J Ethnopharmacol. 1999 Dec 15;68(1-3):3-37. doi: 10.1016/s0378-8741(99)00085-9. PMID: 10624859.
- Riley, J.C.,Willemot,C., Thompson, J.E., *"Lipoxygenase and hydroperoxide lyase activities in ripening tomato fruit".* Postharvest Biol. Technol.1996; 7:97–107.10.1016/0925-5214(95)00032-1.
- Savatović,M., Aleksandra, Tepić Horecki, Šumić, M., and Nikolić, S., *"Antioxidant activity of polyphenol-enriched apple juice."* Acta Periodica Technologica. (2009). 10.2298/APT0940095S.
- Spilimbergo, S., Elvassore, N.,. Bertucco, A., *"Microbial inactivation by high-pressure",* J. Supercrit. Fluids, 22 pp. 55-63, (2002).
- Vasantha Ranasinghe, H.P., and Yu, Li Juan , *"Emerging Preservation Methods for Fruit Juices and Beverages,"* Food Additive, April 7th 2011, Reviewed: August 22nd 2011, Published: February 22nd 2012, DOI: 10.5772/32148.
- Vercet Antonio, Sánchez Gimeno Ana, Burgos Justino, Montañés Luis, López Buesa Pascual. *"The effects of manothermosonication on tomato pectic enzymes and tomato paste rheological properties".* Journal of Food Engineering. 53. 273-278. (2002). 10.1016/S0260-8774(01)00165-0.
- Whitaker, J.R., *"Principles of Enzymology for the Food Sciences",* (2nd edition.), Marcel Dekker, New York (1994).
- www.sugarindia.com.
- Zhao, Wenting, Sun, Yeting, Ma, Yue , Zhao, Xiaoyan. *"Dense phase carbon dioxide treatment of tomato juice: effect on physico-chemical properties, phenolic composition, lycopene isomerisation and in-vitro bioaccessibility."* International Journal of Food Science and Technology, December 2018.

❒

8

Mechanism of Sterilization and Preservation Using Supercritical CO_2

8.1 Bactericidal Effect of CO_2

There has been an ever-increasing need to provide appropriate processes for treatment of food materials to enhance their shelf life (for enhanced storage time without refrigeration) in packed and unpacked state without substantially losing their organoleptic characteristics. For past decades it has been popularly known that gaseous CO_2 can inhibit microbial growth, leading to its use in the preservation of packed foods, although its inactivation effect seems reversible. Several researchers have demonstrated that gaseous CO_2, even at subcritical low pressure is able to inhibit the growth of microorganisms including certain spores and enhance the inactivation rate during irradiation or thermal treatment. Even at pressure as low as 6 bar, this gas exhibits a significant bactericidal or bacteriostatic effect *[Cuq et al., 1993]*. For example, treatment with gaseous CO_2 at 6 bar and 50–55°C has the same lethal effect on several bacteria, fungi, and yeasts as treatment with air at 60–65°C and thus can reduce 50% of the time of pasteurization at a given temperature.

It has been clearly demonstrated that the several microorganisms are irreversibly inactivated in $scCO_2$ (at 35°C ,74 bar) in few minutes and that the bactericidal effect of CO_2 cannot be merely attributed to hydrostatic pressure in the range of 1-100 MPa, but the very efficient cell inactivation that is possible,

M. Mukhopadhayay and A. Chatterjee, *Sterilization and Preservation*,
https://doi.org/10.1007/978-3-031-17370-7_8

is also due to specific biological interactions between CO_2 and microorganisms. Moreover, this specific bactericidal effect is definitely supported by high hydrostatic pressure with and without carbon dioxide. For example, the inactivation of *Escherichia coli* in scCO_2 at 150 bar and 35°C during 15 min was found similar to the one observed at 3000 bar at ambient temperature during the same period of time *[Smelt and Rijke, 1992].* This bactericidal effect is demonstrated to be caused by specific interactions between the living cell and the fluid inside the cell in which CO_2 is readily dissolved.

The microbial inactivation by scCO_2 processing is proven to be superior to that by using subcritical gaseous CO_2 due to its enhanced solubility and mass transfer for dissolution in the fluid inside the cell and membrane-lipids interaction. ScCO_2 takes the advantages of its unique characteristics as mentioned in Chapter 5. On one hand, scCO_2 possesses a good solvation power because of its high density, on the other hand, the high diffusivity and low surface tension of scCO_2 lead to the diffusion of CO_2 into the most intricate parts of the system. The studies conducted so far indicate a positive bactericidal effect of scCO_2 and that the scCO_2-based technology can be used as an alternate method for sterilization.

The efficacy of scCO_2 in sterilization and preservation results from a number of unrelated phenomena. Processing with scCO_2 for sterilization does not require use of high temperatures though the efficacy rapidly decreases at a temperature less than 40°C. Hence nutritional value remains intact in case of foods and heat-sensitive substrates. The presence of water drastically increases the bactericidal effect of CO_2, probably due to the pH effect. An acidic pH seems to favor inactivation of most enzymes as it was demonstrated that there is a strong and irreversible effect of moisture *[Perrut, 2012].* A very high pressure (>200 bar) does not necessarily lead to a significant improvement in efficacy and a medium-pressure in the range (80–150 bar) seems to be sufficient. Moreover, a fast pressurization-depressurization cycling, termed as *process intensity,* is found to increase the inactivation rate *[Chakraborty, 2006].*

8.2 Mechanism for Inactivation of Vegetative Microorganisms

Over the years, researchers have attempted to understand the underlying mechanisms of microbial inactivation behind the bactericidal effect of scCO_2. There are two proposed mechanisms hypothesized for inactivation of the vegetative cells: (i) cell wall rupture and/or (ii) physiological inactivation.

8.2.1 Cell Wall Rupture

It is postulated that $scCO_2$ exerts high pressure on the cells, enabling rapid diffusion of $scCO_2$ into the cells. The cells burst due to volume expansion occurring when high-pressure CO_2 dissolves into the cell fluid. The rupture of cell walls is further assisted during quick depressurization when CO_2 which has entered the bacterial cell, exerts a force in the opposite direction on the cell wall. A physical stress is developed as a result of the pressure difference across the cell wall resulting in the cell wall rupture especially in case of rapid depressurization. The physical stress exerted on the soft tissues causes cell deformation and the cell structure is also weakened due to a strong interaction of the fluid with the lipids (mainly phospho-lipids).

It has been established that rapid pressurization/ depressurization cycles of $scCO_2$ cause membrane disruption and cell lysis. As a result, pressure cycling is a very effective parameter in improving the sterilization efficacy, and may considerably reduce the treatment time for obtaining a given degree of inactivation *[Mukhopadhtay and Chakraborty, 2004]*. Another support for this mechanism comes from the synergetic effect of a very brief pulsed electric field pre-treatment combined with a classical high-pressure $scCO_2$ treatment with pressure cycling that leads to a very significant improvement of the sterilization efficacy *[Spilimbergo, et al., 2013]*, as the pre-treatment may render the cell membrane more fragile (or rupture it). In *E. coli* and *Staphylococcus aureus*, the inactivation rate jumps from 2.5 to 8.5 in log reduction and from 3.5 to 7.8 in log reduction respectively, upon processing with $scCO_2$ only or combining $scCO_2$ processing with such pre-treatment at the same conditions.

As can be seen from *Figure 8.1* illustrating the ESEM pictures of the untreated and $scCO_2$ treated *Bacillus* cells, the treated cells are deformed and swollen as compared to the untreated cells. This provides the evidences that cell rupture happens during pressurization. This validates the fact that the dissolution of $scCO_2$ in the cellular fluid swells the cell volume beyond its elastic limit, which also results weakening of the structure of the cells. An increase in the intensity of the processing protocol causes weakening of the internal organization of the cell and ultimately disrupts the cell membranes due to which the internal structure collapses resulting in inactivation *[Chakraborty, 2006]*.

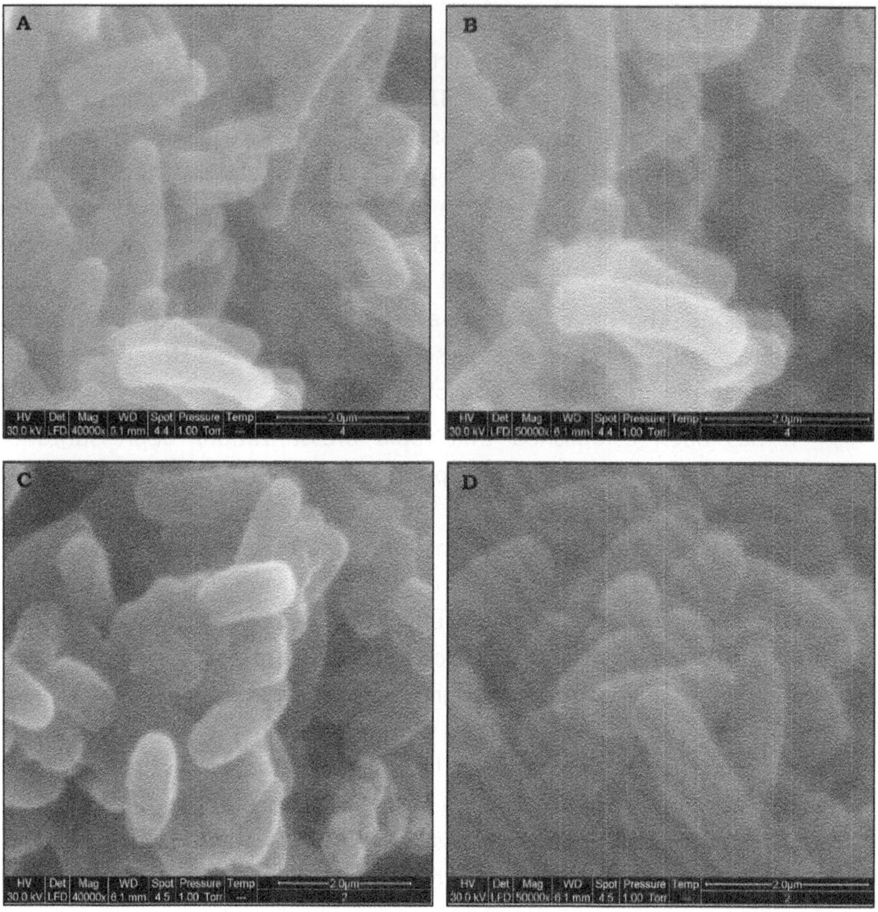

Fig. 8.1: ESEM pictures of Bacillus cells, A and B are untreated cells, C and D are $scCO_2$-treated Cells at 40,000 and 50,000 magnification respectively. *[Chakraborty, 2006]*

8.2.2 Physiological Inactivation

The physiological inactivation of microorganisms is primarily attributed to the development of physical, chemical, and biological stresses during the processing protocol followed using treatment with $scCO_2$. The combination of these stresses facilitates sterilization of both aerobic and anaerobic microorganisms, molds, and fungi. Physiological inactivation is a complex mechanism that can be described by the simultaneous occurrence of seven steps *[Soares et al., 2019]*.

Step 1: Dissolution of CO_2 lowering pH

Step1 is described as the dissolution of pressurized CO_2 in the cell fluid. Due to enhanced mass transfer of scCO_2, owing to its excellent transport properties and the processing protocol, the amount of CO_2 entering the cell causes chemical stress. The dissolution of CO_2 in water present in the cell fluid causes a reduction in pH due to the formation of carbonic acid (H_2CO_3).

Table 8.1 presents compiled data of CO_2 solubility as a function of pressure and temperature. It can be seen that the solubility of water in CO_2 increases with pressure and decreases with temperature. CO_2 dissolution in water results formation of carbonic acid. As a result, water in contact with CO_2 becomes acidic.

Table 8.1: Solubility of CO_2 as a function of pressure and temperature. [*Spilimbergo et al.,* 2002]

Temperature (°C)	Pressure (bar)	Solubility (mol/kg)
20	50	1.44
	100	1.49
	150	1.53
	200	1.59
30	50	1.3
	100	1.35
	200	1.55
40	50	1.2
	100	1.15
	200	1.43
50	50	1.13
75	100	0.9
	200	1.15

The carbonic acid formed due to dissolution of CO_2 during processing interacts with the intracellular ions, such as, calcium, resulting in its precipitation and subsequently affects the normal cellular functioning. Further, dissociation of carbonic acid causes formation of hydrogen cations (H+), lowering the pH as given by the following chemical equations:

$$CO_2(g) + H_2O(aq) \rightarrow H_2CO_3(aq) \ldots\ldots\ldots Ka = 4.2*10^{-7}$$

$$H_2CO_3(aq) \rightarrow H^+(aq) + HCO_3^-(aq) \ldots\ldots Ka = 4.8*10^{-11}$$

$$Ca^{2+}(aq) + 2HCO_3^-(aq) \rightarrow H_2O(g) + CO_2(g) + CaCO_3(s)$$

As can be seen from *Figure 8.2*, pH decreases with increasing pressure owing to an increase in dissolution of CO_2 and subsequently resulting formation and dissociation of carbonic acid. For example, pH reduces to 3.1 in water at 10MPa at 38°C *[Spilimbergo et al., 2002]*.

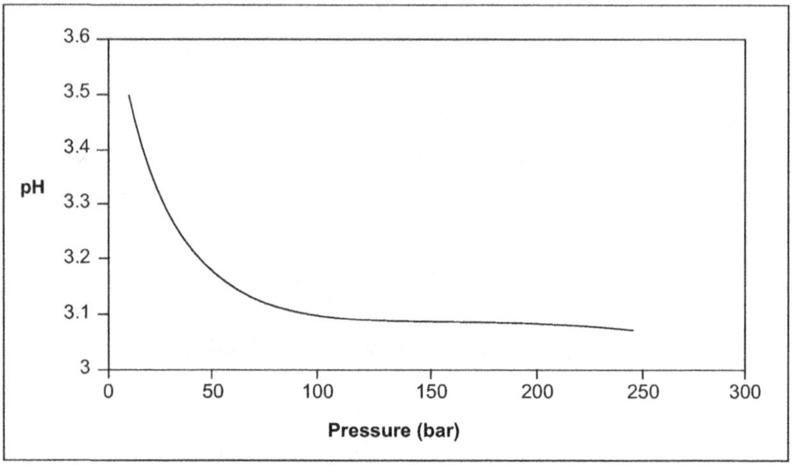

Fig. 8.2: Extracellular pH as a Function of Pressure at 38°C *[Spilimbergo et.al., 2002]*

The drop in pH during processing is substantiated by a lower value of pH (than the original) observed in the treated samples of water, milk, and tomato juice after processing with $scCO_2$ due to the presence of some residual dissolved CO_2, as shown in *Figure 8.3*.

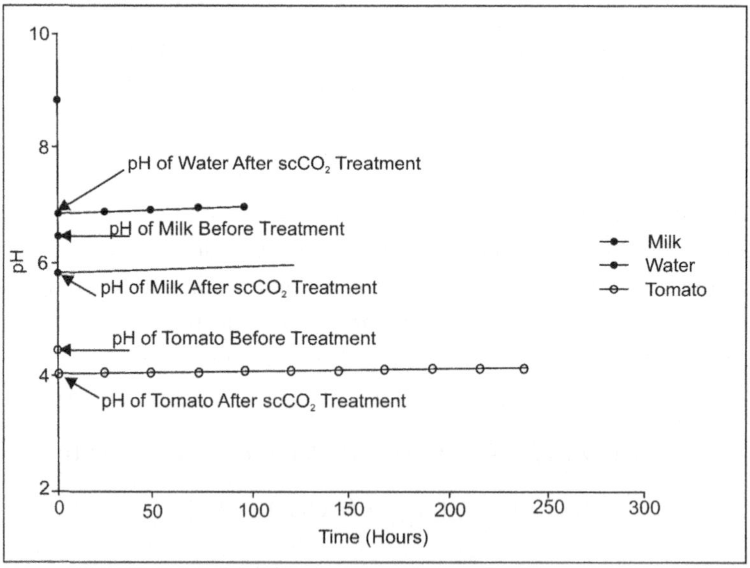

Fig. 8.3: Presence of residual CO_2 in the food systems substantiated by the pH change in the treated samples over a period of time *[Chakraborty, 2006]*

This lowering of pH of the food system by $scCO_2$ during processing plays a key role in inhibiting microbes and enzymes. A lower pH is able to reduce the microbial resistance to inactivation due to biological stress. The dissolved CO_2 reduces the surface tension and viscosity of the juice, which may render evaporation of some amount of moisture from the system. As a result, there is continual moisture loss from the cell and an irreversible precipitation of the dissolved nutrients may take place within the cell, which may again lead to their weakening.

However, there are certain microorganisms, despite tolerating low pH, which can also be inactivated rapidly when these are exposed to $scCO_2$ due to other effects, as mentioned earlier.

Step2: Modification of cell membrane

Step 2 relates to modifications of cell membrane due to dissolution of CO_2. As CO_2 has high affinity to plasma membrane, CO_2 can diffuse into the cellular membrane and accumulate into its lipophilic inner layer, causing structural and functional disorders of the cell membrane *[Garcia-Gonzalez et al., 2007]*. This phenomenon has been attributed to the loss of the lipid chain, leading to increased fluidity. CO_2 is likely to contribute to an increase in permeability of the cell membrane, thus simultaneously allowing more penetration of CO_2 through the cell membrane and excess CO_2 can accumulate in the cytoplasm of the vegetative cells. This leads to the third step.

Step3: Disturbance of homeostasis due to pH reduction

It has been demonstrated that dissolution of CO_2 leads to lowering of external pH, which in turn leads to lowering of internal pH in presence of CO_2 *[Spilimbergo et al., 2002]*. This reduction in pH disturbs homeostasis of the cells leading to their inactivation as explained in subsequent steps.

Step 4: Inactivation of enzymes due to pH reduction

Enzyme activity largely depends on internal pH. When cytoplasm becomes too acidic, the low internal pH is likely to inhibit or inactivate the key enzymes like *decarboxylases* which are essential for metabolic and regulating processes. Excess of CO_2 thus inhibits cellular metabolism by breaking the metabolic chain. It has been reported that most of the enzymes that participate in the metabolism of carbohydrate and amino acid, function at a neutral range of pH and their activity declines sharply outside this range. If their pH deviates to higher or lower values, intensive electrostatic attraction and repulsion occur between charged side chains of proteins. This eventually causes the folding and unfolding of peptide chains resulting in denaturation of enzyme proteins in the vicinity of pH 3-4. Thus lowering of pH by $scCO_2$ treatment is speculated to be underlying mechanism for the enzyme inactivation.

Yao et al. [2014] also suggested that if intracellular pH was low enough, denaturation of DNA in *E. coli* would also happen and the double stranded helix of DNA would unwind. This is attributed to the enzymatic reactions and enzyme instability in $scCO_2$ *[Ishikawa et al., 1995; Perrut,1994].*

Step 5: Inhibitory effect of molecular CO_2 and HCO_3^- on metabolism

Step 5 is a direct inhibitory effect of two molecules CO_2 and HCO_3^- on microbial metabolism which presumably depends on their concentrations in the cytoplasm, as these molecules directly affect carboxylation and decarboxylation reactions *[Mukhopadhyay and Chakraborty, 2004].*

Step 6: Disturbance of electrolyte balance

This involves a disordering effect on the intracellular electrolyte balance. An accumulation of CO_2 inside of the cytoplasm of a cell raises the pressure and, when released, it converts HCO_3^- to carbonate (CO_3^{2-}). This anion could precipitate intracellular calcium, magnesium, and similar cations from cells and cell membranes, disordering the electrolyte balance and producing lethal damage to the biological system of the cells *[Soares et al., 2019].*

Step 7: Removal of vital constituents from cells

It involves the removal of vital constituents from cells and cell membranes. $ScCO_2$ is known to have a high solvating power and so it may extract intracellular lipid constituents, such as, phospholipids and hydrophobic compounds. This lipid extraction disturbs the cellular biological system, thus promoting inactivation. It has been shown that there is a direct relationship between the leakage of nucleic acid, potassium, and magnesium cations and the inactivation of *E. coli* treated by $scCO_2$ *[Mukhopadhyay and Chakraborty, 2004].*

8.2.3 Biological stress

During processing with $scCO_2$, the oxygen present in the vessel as well as the dissolved oxygen is removed. Hence oxygen is not available to the cells, which is required for their normal functioning. In addition, the lack of oxygen (due to presence of CO_2) changes the biological environment required by the cells to survive. This causes biological stress in the cell, blocking its normal metabolic pathway and ultimately causes cell death. This is substantiated by the fact that for certain systems the inactivation of microorganisms persists even after the processing, till a minimum microbial count is reached. After the processing, though most of the CO_2 diffuses out, the presence of even small residual CO_2 and the resulting residual stress within the cells slowly inactivates the microorganisms till all CO_2 diffuses out.

8.3 Inactivation Mechanism of Bacterial Spores

It is interesting to note that resistance to inactivation offered by most spores is much stronger than for vegetative cells and that they survive when contacted with scCO_2 at mild temperatures (say <60°C), even at a very high pressure (2500 bar) for certain strains. However, it was established that a fast pressurization-depressurization cycling a number of times improves the inactivation rate *[Mukhopadhyay and Chakraborty, 2005]*. On the other hand, it is reported that once a contaminated material is sterilized with scCO_2, then contacting the treated material with irradiation further improves the sterilization efficiency. It is observed that temperature is the key-parameter for spore inactivation, which is drastically improved at a temperature beyond 70°C depending on the strain. Most spores including the most resistant ones (such as, the heat-resistant *G. stearothermophilus*) can be inactivated by a combination of two or more types of treatment techniques, such as, lowering pH or addition of a strong oxidant to scCO_2, such as, hydrogen peroxide(H_2O_2), tert-tributyl hydroperoxide (t-TBHP) or a mixture of both. Even addition of peracetic acid or trifluoroacetic acid even at very low concentration, permits to reach a high efficacy at mild temperature.

The mechanism for inactivation of bacterial spores is different due to a great resistance exerted by their inherent unique structure and is mostly attributed to a structural change of spore structures. A high-pressure CO_2 treatment seems to induce structural changes in endospores, specifically in the inner membrane of spores. This lethal effect is explained in terms of a two-step mechanism. As explained earlier, treatment with scCO_2 increases the permeability of the inner membrane of endospores, resulting damage of crucial proteins associated with spore germination (located in the inner membrane), thus inhibiting the endospore germination. Analysis of the structure of bacterial spores by Scanning Electron Microscopy (SEM) confirmed a difference in the structure of the residual spores before and after treatment with scCO_2. The spore inactivation seems to be caused by disruption of the spore's outer layers, especially in the presence of a strong oxidant additive in scCO_2. This was confirmed by the fact that spores processed with scCO_2 + H_2O_2 show a less distinctive envelop, both for external and internal layers, which correlates with the high killing rate of spores *[Soares et al., 2019]*.

8.4 Mechanism of Virus inactivation

In a pioneering work, *Stahl et al. [1988]* had reported a weak inactivation (~2 log reduction) of *Coliphage* virus by $scCO_2$ treatment at a very high pressure of 2500 bar. Subsequently complete inactivation of the same virus type (e.g., *E. coli* T1 to T5 phages) was claimed to be achieved by $scCO_2$ at 300 bar and 50°C during a 30-min treatment time in an aqueous medium with $scCO_2$-dispersed "micro-bubbles". However, decrease in pH during treatment with $scCO_2$ in presence of water has a drawback in that it leads to irreversible denaturation of pH-sensitive proteins *[Perrut, 2012]*.

Addition of a strong oxidant, such as, H_2O_2 or PAA, may increase the efficacy of $scCO_2$ treatment at lower temperatures by penetrating into the core and subsequently oxidizing the outer layers of the vital cell structures, as they are slightly affected after $scCO_2$ sterilization. However it is still not clear how cell structures and vital components are damaged by high-pressure CO_2 treatment, so the detailed inactivation mechanism is not yet fully unearthed.

8.5 Mechanism of Inactivation of Enzymes

Preservation of solid and liquid foods using $scCO_2$ is aimed at extending shelf life not only by inhibiting microorganisms, but also inactivating the deleterious enzymes present in them. For several years $scCO_2$ treatment has been demonstrated to be effective for inactivation of indigenous enzymes in foods, such as, *polyphenol oxidase* (*PPO*), *pectin methylesterase* (*PME*), and *lypoxygenase* (*LOX*) which are responsible for causing unacceptable deterioration, such as, the change in color, texture, flavor, and taste over a period of time. The probable mechanisms in effectively inactivating these enzymes by the $scCO_2$ treatment are attributed to several factors, such as, effects of pressure, temperature, interaction of CO_2 molecules, and the process intensification due to pressurization-depressurization cycling.

8.5.1 Inhibitory Effect of Molecular CO_2

The enzyme inactivation is mostly attributed to the drop in pH as well as to the processing protocol, as discussed earlier. Over and above the pH lowering effect in the $scCO_2$ treatment, an additional mechanism proposed for enzyme inactivation is due to the inhibitory effect of contact of molecular CO_2 with the target enzyme leading to the formation of a complex with CO_2 molecules or decomposition of the enzyme. It was found that protein-bound arginine can easily interact with CO_2 at low pH to form a bicarbonate complex *[Weder et al., 1992]*. Subsequently it was reported that CO_2 in excess may form covalent carbamates with free amino groups on the surface of the enzyme *[Fricks et al., 2006]*. These carbamates are able to remove charge from the histidine and lysine residues to contribute to the loss of enzyme activity.

Another factor which could also contribute to enzyme inactivation is cited as the effect of hydrophobic interactions between CO_2 and the active sites of the enzyme. Some enzymes may be decomposed by CO_2 due to the pressure-induced effect of $scCO_2$ treatment which could cause changes in protein backbone structure leading to its dissociation. Acid *protease* was reported to be inactivated by micro-bubbled $scCO_2$ due to this irreversible decomposition of the protein structure *[Yoshimura et al., 2006]*. CO_2 molecules can thus perturb the subtle balance of intramolecular and $scCO_2$-protein interactions, leading to complete unfolding of the enzyme polypeptide chain, resulting in a loss of their redox active sites and the site-specificity of the specific enzyme. The enzymes thus undergo irreversible denaturation in the environment of $scCO_2$, leading to its selective inactivation *[Hu et al., 2013]*.

8.5.2 Effect of Characteristics of $ScCO_2$

In addition, owing to zero surface tension of $scCO_2$ and ultimate collapse of the active site of enzymes, $scCO_2$ can easily penetrate into the large enzyme molecules. Due to its good solvent capacity it could also strip essential water molecules from the enzyme microenvironment, with the water partitioning between the enzymatic support and the $scCO_2$ solvent, thus causing enzyme inactivation.

8.5.3 Effect of Pressurization-Depressurization Cycling

Finally, as reported by *Hu et al. [2013]*, the inactivation mechanism by pressurization-depressurization cycling may be attributed to two steps, namely, formation of a gas-liquid interface by CO_2 bubbles and their fast escape at the end of the pressure cycle. Owing to their amphipathic nature, enzyme proteins are likely to spontaneously move to the gas-liquid interface to attain a relatively low Gibbs free energy state where protein chains would unfold due to orientation of nonpolar residues to the gas phase and polar residues to the liquid phase. The unfolded molecule would then undergo aggregation in its hydrophobic regions, resulting in increased protein concentration at the gas-liquid interfacial region. The aggregates eventually desorb from the gas-liquid surface and precipitate above a certain concentration in the solution. Sudden release of CO_2 pressure results in rapid gas expansion from the enzymes and can induce a more drastic expansion and thus decompose the aggregated enzymes into smaller particles. Alternatively, the loss of enzyme activity observed following depressurization can be explained by the rapid release of the CO_2 dissolved in the bound water around the protein *[Hu et al., 2013]*.

References:

- Chakraborty Anuradha, *Sterilization and Stabilization of Food Products Using Supercritical Carbon Dioxide,* Ph.D. Dissertation, Indian Institute of Technology Bombay, 2006.
- Cuq, J.L.; Roussel, H.; Vivier, D; Caron, J.P., *"Effects of gases under pressure: influence on the thermal inactivation of microorganisms",* Sciences des Aliments, 13, 677–698, (1993).
- Fricks, Alini T., Souza Dayse, P.B., Oestreicher, Enrique, Guillermo, Antunes, Octavio, A.C., Girardi, Jonathan, Oliveira, Débora, S., Dariva, Cláudio, *"Evaluation of radish (Raphanus sativus L.) peroxidase activity after high-pressure treatment with carbon dioxide,"* The Journal of Supercritical Fluids, Vol 38, Issue 3, Pages 347-353, 2006.
- Garcia-Gonzalez, L.; Geeraerd, A.H., Spilimbergo, S. Elst, K., Van Ginneken, L., Debevere, J., Van Impe, J.F., Devlieghere, F., *"High pressure carbon dioxide inactivation of microorganisms in foods: the past, the present and the future",* Int. J. Food Microbiol. 117 (1) 1–28, (2007).
- Holmes, Justin, Lyons, Daniel, Ziegler, Kirk, *"Supercritical Fluid Synthesis of Metal and Semiconductor Nanomaterials,"* Chemistry (Weinheim an der Bergstrasse, Germany). 9. 2145-51. 10.1002/chem.200204521, 2003.
- Hu, Wanfeng; Zhou, Linyan; Xu, Zhenzhenl Zhang, Yan; Liao, Xiaojun, *"Enzyme Inactivation in Food Processing using High Pressure Carbon Dioxide Technology,"* Critical Reviews in Food Science and Nutrition, 53:145–161, 2013.
- Ishikawa, H., Shimoda, M.; Kawano, T.; Osajima, Y.; *"Inactivation of enzymes in an aqueous solution by micro-bubbles of supercritical carbon dioxide",* Bioscience Biotechnology Biochemistry, 59 628–631, 1995.
- Mukhopadhyay, Mamata and Chakraborty, Anuradha, *"Process for Sterilization of Biomaterials Using Supercritical Fluids",* Indian Patent no.211305, 2007, Application no. 543/MUM/2004.
- Mukhopadhyay, Mamata and Chakraborty, Anuradha, *"Lipoxygenase-Inactivated and Sterilised Legumes and Cereal Products"* Indian Patent no.282732, 2017, Application no. 540 /MUM/2005.
- Perrut, M.; *"Enzymatic reactions and cell behaviour in supercritical fluids",* Chemical Biochemical Engineering Quarterly 8 25–30, (1994).
- Perrut Michel, *"Sterilization and virus inactivation by supercritical fluids (a review),"* The Journal of Supercritical Fluids, Volume 66, Pages 359-371, June 2012.
- Smelt, J.P.P.M.; Rijke, G.F *"High pressure treatment as a tool for pasteurization of foods",* in: C. Balny, et al. (Eds.), *"High Pressure and Biotechnology",* vol. 224, John Libbey Euro text Ltd., pp. 461–463, 1992.
- Soares Goncalo C., Learmonth, David A., Vallejo Mariana, C., Davila Sara Perez, Gonzalez Pio, Sousa Rui A., Oliveira Ana L., *"Review Supercritical CO2 technology: The next standard sterilization technique,"* Materials Science &

Engineering C 99, 520–540, 2019.

- Spilimbergo, S.; Dehghani, F.; Bertucco, A.; Foster, N.R; *"Inactivation of bacteria and spores by pulse electric field and high-pressure CO_2 at low temperature, Biotechnology & Bioengineering,"* 82, pp.118–125, (2003).
- Spilimbergo, S., Elvassore, N.; Bertucco A., *"Microbial inactivation by high-pressure,"* J. Supercrit. Fluids, 22, pp. 55-63, (2002).
- Stahl E., Quirin KW., Gerard D., *"Non-Extractive Applications. In: Dense Gases for Extraction and Refining,"* Springer, Berlin, Heidelberg, 1988.
- Weder, Jürgen;., Bokor, K.P; Maria, V., Hegarty Mervyn, P., *"Effect of supercritical carbon dioxide on arginine,"* Food Chemistry, Volume 44, Issue 4, Pages 287-290, 1992.
- Wu Yin, Yao Shan-Jing, Guan Yi-Xin, *"Inactivation of Microorganisms in Carbon Dioxide at Elevated Pressure and Ambient Temperature,"* Ind. Eng. Chem. Res., 46, 19, 6345–6352, 2007.
- Yao, C.; Li, X.; Bi, W.; Jiang, C., *"Relationship between membrane damage, leakage of intracellular compounds, and inactivation of Escherichia coli treated by pressurized CO_2,* J. Basic Microbiol, 54 (8), pp. 858–865, (2014).
- Yoshimura,T.; M. Shimoda., H. Ishikawa , M. Miyake I. Hayakawa, K. Matsumoto Y. Osajima, *"Inactivation Kinetics of Enzymes by Using Continuous Treatment with Microbubbles of Supercritical Carbon Dioxide",* Journal of Food Science, Volume 66, Issue 5, pp. 694-697 2006.

❐

9

Sterilization of Biomaterials and Medical Devices with Supercritical CO_2

The use of biomaterials and medical devices has increased manifolds in the last decade to an extent that it is now an integral part of many treatments. Although, this has been a boon in more ways, but it has also led to the development of new families of diseases that are directly related to the use of the biomaterials and medical devices. The grave issue that medical professionals face is the threat of microbial contamination of the devices often causing devastating infections leading to the loss of implanted devices.

9.1 Safety and Success of Medical Devices

Biomaterials are essentially natural or synthetic materials that are used in medical devices suitable for inclusion in biological systems for medical purpose (therapeutic or diagnostic). It is crucial for the safety and success of a medical device, that it is biologically inert, free from pathogens, and does not trigger an adverse reaction from the body. The infection may occur owing to an external source, i.e., the environmental conditions and response of the host. Healthcare-associated infections are a major concern and hinder the success of a treatment. The consequences of nosocomial infections not only pose a threat to the patient's health but also impose an additional economic burden. Generally, the approach adapted to mitigate and minimize the risk of such infections, includes a combination of aseptic practices and an appropriate treatment of the medical devices.

M. Mukhopadhayay and A. Chatterjee, *Sterilization and Preservation*,
https://doi.org/10.1007/978-3-031-17370-7_9

The basic purpose of biomaterials is its applicability in the medical industry. Biomaterials are used as implants or as medical devices and come into human contact. It is thus inevitable that biomaterials will be in contact with human tissues, and must first be sterilized and maintained sterile with certainty until usage to ensure success. Sterility of materials is usually assured by less than one in one million devices having a microbial presence. It is imperative that these biomaterials do not act as carriers of contamination or diseases, do not elicit antagonistic reaction from the body, and are essentially inert. This is where sterilization of biomaterials plays a key role, apart from cleaning and disinfection.

Biomaterials are sensitive to heat and harsh treatment conditions, thereby making it essential to use a benign sterilization process. The process of sterilization has to be such that it ensures the functionality of the material to remain intact and sterile starting from the manufacturing plant to the clinic/ hospital. This is the key to the success of the medical devices and implants. Thus, microbial contamination is an important issue in the manufacture of pharmaceutical and medical devices, which directly affects the patients' safety and also hinders the stability of the drugs and formulations. In this regard, sterilization serves the purpose of total inactivation of viable microorganisms, including vegetative and sporulated bacteria, fungi, yeasts, and viruses. The effect of sterilization, the possibility of modifications of the surface of the biomaterial, and its characteristics need to be assessed for the success of the biomaterial post the treatment procedure.

9.2 Methods of Contamination and Infectious agents

The main contaminants in medical devices and biomaterials are viruses, fungi, protozoas, and bacteria. Researchers worldwide largely opine that the water pipelines in surgical and dental units may be the potential transmission pathway of the *hepatitis* virus as well as *Pseudomonas aeruginosa* and *Legionella*. Fungi, such as, *Candida albicans* and *Aspergillus* are reported to be related to the early onset of prosthetic valve endocarditis just after implantation. The protozoan organism *Acanthamoeba* is reported to be a rare but devastating cause of keratitis leading to loss of vision when it colonizes on contact lenses *[Moore et al.,1985]*.

The microorganisms that are involved in biomaterial contamination have low virulence in planktonic form. However, they can cause serious infection if they form a biofilm. A cluster of microorganisms adhering to a surface glued in a matrix referred to as a biofilm. This cluster can be made up of either homogenous or heterogenous population. Biofilm constitutes a protected mode for the bacteria to survive in a hostile environment with respect to planktonic forms. In particular, it allows bacteria to optimize nutrient uptake, to be sheltered from the removal forces, and to be protected from the host defence mechanisms and from the potential toxic or harmful substances of the surrounding environment *[Rimondini et al., 2005].*

The most significant hospital-acquired infections, based on frequency and potential severity, are those related to procedures, e.g., surgical site infections and medical devices, including urinary tract infection in catheterized patients, pneumonia in patients intubated on a ventilator, and bacteraemia related to intravascular catheter use. At least half of all cases of nosocomial infections are associated with medical devices. Such devices may be utilized only for short time intermittently or for months and years, or permanently. Plastic devices are easily colonized with bacteria and fungi, which are able to be colonized by microorganisms at a rate of 0.5 cm per hour. A thick biofilm is formed within 24 hours on the entire surface of these plastic devices once inoculated even with a small initial number of bacteria *[Guggenbichler et al., 2011].*

9.3 Factors Involved in Sterilization of Biomaterials

Selection of an effective sterilization process is dependent on several factors, which include the state of the biomaterial (liquid, solid or semi-solid), its sensitivity to heat, moisture, chemical, and radiation. The natural bioburden of biomaterials and devices plays a key role in determining the selection of a sterilization process. The higher the initial natural bioburden level, the severe is the sterilization requirement. It is worth noting that the contaminating microflora will primarily consist of vegetative bacteria, but the presence of spores which are more resistant to sterilization needs to be checked. Bacterial spores (or endospores) are basically a survival mechanism they have adapted to the highly differentiated dormant cells. Spores are heat resistant and resistant to harsh chemicals, free radicals, and radiation. All of these make it difficult to inactivate spores, thus, enabling them to tide over the unfavorable growth conditions. Many fungi also have the capability of sporulation, mostly as part of sexual or asexual reproduction cycle.

One of the most important criteria for sterilization process to be successful is the inactivation of spores. Thus, in accordance to the FDA requirement, it is necessary to test for presence or absence of spores, which will be indicative of successful sterilization. The second factor is the permeability of the sterilizing agent or the sterilant. The sterilant (gas or liquid) must have the ability to come in direct contact with the microorganisms during the sterilization process. The presence of biofilms on the biomaterial and designs of the medical device that limit the contact will reduce the efficacy of sterilization by impairing exposure of the sterilant to the microbial cells. Also, biomaterials must be compatible with the sterilization processes. The presence of certain components, such as, salt and organic residues, may compromise the efficacy of sterilization. Another important factor is that the sterilization process should not alter material properties and stability that are critical to functioning of the device *[Qiu et al., 2011]*.

9.4 Current Methods of Sterilization and Disinfection

There are three ways to ensure sterilization of a product: (1) overkill method, (2) combined biologic indicator–bio-burden method, and (3) bio-burden method. Of these the first one is the most widely utilized technique. Biomaterials of biologic origin that are widely used include allografts and biological material derivatives. For the success of the allograft application, there are a few critical factors that play a major role. These include sterility, reduction of antigenicity, and preservation of the natural biological and biomechanical properties *[Qiu et al., 2011]*. Numerous strategies are being adopted to prevent bacterial contamination. *Table 9.1* lists these strategies *[Rimondini et al., 2005]*.

Table 9.1: Common strategies to prevent microbial contamination *[Rimondini et al., 2005]*

Devices	Strategy
Dialysis Catheters PVC endotracheal tubes Silicone intracular lenses Long term catheters Dental implants	Surface treatment
Urological devices	Surface modification
Orthopaedic soft tissue implant Cental venous catheters	Core material
Bone cement	Drug delivery system
Generic Biomaterials	Inhibition of gram negative adhesion

Sterilization is the final step in the manufacture of any biomaterial, which is expected to kill any microorganisms that may have been present, rendering the biomaterial sterile. This is an important step since the future application of the biomaterial and the subsequent consequences on the human body depend largely on the success of sterilization.

Processes and technologies developed to obtain inactivation of microorganisms are based on physical or chemical agents. However, they may be used in a synergic combination of two or more treatments. The main aim of doing this is to inhibit the cell activities involved in metabolism or genetic transcription, inducing cell death, stopping the replication ability of contaminants. The Regulatory bodies have classifications for medical device risks, that account for the global risk associated with the device including the infection risk, malfunctioning risk, electrical risk, and personnel risk. For sterilization purposes, the Spaulding classification, based on infective risk, is pertinent and is presented in *Table 9.2 [Tessarolo and Nollo, 2008]*.

Table 9.2: Infective risk classification of medical devices according to Spaulding *[Tessarolo and Nollo,2008]*

Application	Level of Risk	Spaulding classification	Recommendation
In a close contact with a break in the skin or mucous membrane	High	Critical	Sterilization
For introduction into sterile body areas in contact with intact mucous membrane	Intermediate	Semi critical	High level disinfection or sterilization
Prior to use in immune compromised patients in contact with healthy skin	Low	Non-critical	Low level disinfection

There are quite a few sterilization technologies that are being used for sterilizing biomaterials. Sterilization of biomaterials is most commonly achieved using radiation, chemical or thermal methods. The use of ionizing radiation and ethylene oxide offers advantages for scale-up and are typically preferred as methods for industrial level sterilization. Autoclave sterilization and chemical surface treatments are most commonly used for disinfection of glass and metallic devices. While patient safety should be the primary consideration in the choice of appropriate methods to ensure sterility, attention should also be paid to the impact of the sterilization method used on the biomaterial and any potential changes in functional performance.

Application of heat or pressure or both are commonly used sterilizing principles. The dry heat-sterilization destroys microorganisms by destructive oxidation of cell components. To achieve killing of resistant spores by dry heat, usually requires a temperature of about 160°C for 2 h. Thus, dry heat sterilization cannot be used for heat sensitive biomaterials. Steam sterilization or moist heat-sterilization is widely used. It acts by irreversible coagulation and denaturation of enzymes and structural proteins in the microorganisms. It generally requires 121°C for 15–30 min to kill resistant spores. It is therefore unsuitable for moisture-and heat-sensitive biomaterials.

UV rays are usually used to control airborne microorganisms and remove surface contaminants from the production and processing site of biomaterials. UV radiation may be incorporated as a part of a combined strategy to reduce overall bioburden during a manufacturing process, but may not be effective as the sole sterilizer.

Gamma radiation is used as a low-temperature sterilization alternative. These are high energy photons emitted by a ^{60}Co source and have no mass or charge. The wavelengths of gamma rays are less than 10^{-10} m, with frequencies greater than 10^{18} Hz and energies at the 100 keV level. Gamma radiation has greater penetration power; it is safe, cost-effective, and can be used for a wide range of medical products, (e.g., tissues for transplantation, pharmaceuticals, medical devices, disposable medical equipment). Gamma radiation kills microorganisms by attacking the DNA molecule through both direct and indirect mechanisms. Gamma sterilization does not leave any chemical residues and does not impart radioactivity.

The final packaged product of biomaterials and medical devices are irradiated with high energy E-beams. The range of penetration of these electrons is dependent on their energy, the atomic number of the absorbing product, and its density. The mechanism of inactivation of microorganism using an E-beam is similar to that of gamma radiation. One of the advantages for E-beam radiation is that its high energy power can sterilize in seconds.

Ethylene oxide (ETO) is a widely used sterilant. ETO works by alkylating proteins, DNA, and RNA in microorganisms, which prevents normal cellular metabolism and replication, resulting in killing or making the microorganism non-viable. The bioburden reduction increases with the increase of ETO concentration, exposure time, and temperature. The commonly used operational ranges for ETO concentration, temperature, humidity, and exposure time are 450–1200 mg l^{-1}, 29–65 °C, 45–85%, and 2–5 h, respectively. ETO poses hazardous consequences to people and the environment due to its mutagenicity potential and risks associated with handling a flammable gas. Also, ETO leaves residue in the final sterilised product. Despite the toxicity and safety concerns related to ETO, it is the only acceptable sterilization method so far.

Gas plasmas are generated by exciting gas molecules with radio frequency or microwave to produce charged particles (i.e., free radicals). Gas plasma process uses mechanisms combining the action of UV irradiation generated from excited molecules and that of oxygen free radicals. UV irradiation causes lesions of DNA strands of microorganisms. Intrinsic photon desorption resulting from UV photons breaks the chemical bonds of the other microorganism materials to produce volatile by-products (e.g., CO and CHx). Reactive species within a plasma field (atomic and molecular radicals or excited molecules in a metastable state) undergo chemical reactions with cell components to initiate spontaneous etching and thereby disrupt the structure and metabolism of microorganisms, yielding the final oxidation products, such as, CO_2 and H_2O. The type of seed gas is an important variable affecting the effectiveness of the plasma sterilization process *[Qiu et al., 2011].*

9.5 Terminal Sterilization Using $ScCO_2$

Terminal Sterilization is defined as a process wherein the product is subjected to sterilization within its sterile container or barrier system. Terminal sterilization is considered as a part of the entire manufacturing process and is usually carried out at or near the end of the manufacturing process. This entails a very critical and efficient step for the manufacturing of sterile products. An important aspect of terminal sterilization is that it needs immaculate process control and should provide a high assurance of sterility. Terminal Sterilization is routinely applied to a wide variety of implantable medical devices and other medical products that are used during implantation procedures. Combination products with the device as the primary mode of action are sterilised using only terminal sterilization. There are no other better options available currently. The application of terminal sterilization, other than the steam sterilization, in the pharmaceutical industry has been limited due to material compatibility challenges. Terminal sterilization of biologic products using radiation is also limited with the exception of tissue products for tissue banks. As the combination product market expands and evolves, so does the need to expand and evolve the application of terminal sterilization solutions *[Lambert et al., 2011].*

Supercritical CO_2 can be used for sterilization of a varied range of products for microbial inactivation to reach terminal sterilization in several applications, such as, regenerative medicine, food, clinical wastes, drug products, or water treatment. Apart from its efficacy, it is also important that the sterilization with $scCO_2$ treatment should be innocuous to the material without significantly changing its composition and properties. Sterilization with $scCO_2$ has opened up a whole new avenue for medical research on organ transplantation with the preparation of ready-to-use biological grafts. Also, contamination is recurrent in biological tissues owing to its handling. Effective sterilization is the key to prevent disease transmission associated with the use of biological tissue grafts. The existing alternative sterilization techniques are not as effective due to low penetration capacity, the pernicious inactivation of thermo-labile compounds, and the degradation of materials under radiation or oxidative environments resulting in low sterilization efficacy or loss of biological activity *[Ribeiro et al., 2019].*

Supercritical CO_2 is appealing for sterilization due to the ease of attaining the supercritical state, the non-reactive nature, and the ability to readily penetrate substrates. The scCO_2 treatment is a promising technology for sterilization, especially the terminal sterilization of sensitive biomaterials. But inactivating bacterial spores rapidly or within a short time period, has proven a barrier to the use of this technology for effective terminal sterilization *[White et al., 2006]*. However, scCO_2 treatment is a promising strategy for the terminal sterilization of sensitive biomaterials at low temperature, especially in combination with low amounts of additives. Terminal sterilization with such a scCO_2 treatment effectively inactivates microorganisms including bacterial spores.

9.6 Effectiveness of ScCO_2 Treatment of Biomaterial with Select Additives

The most important and advantageous aspect of using scCO_2 is that it is non-reactive, and treatment with scCO_2 does not compromise the morphology, structure, and chemical and mechanical properties of the biomaterials. A wide range of gram-positive and gram-negative vegetative bacteria can be successfully inactivated using scCO_2 treatment. At this point, it is worth noting that the inactivation of bacterial endospores using this technology requires either high temperatures and pressure or prolonged incubation time, as spores by nature are resistant and difficult to kill. These treatment conditions may not be suitable for sensitive biomaterials like the polysaccharides, tissue samples, or proteins.

Adding low amounts of volatile additives during the scCO_2 treatment has demonstrated encouraging results with an increase in the inactivation rate of bacterial endospores. The effect of scCO_2 processing on the mechanical properties of decellularized tissues, especially bone and tendon has been extensively evaluated. A study that was conducted to understand the biomechanical properties of allografts (human tendon and cortical bone), reveals and reports improved biomechanical properties post scCO_2 treatment, as compared to those after gamma irradiation *[Nichols, 2009]*. Furthermore, it has been demonstrated that scCO_2 sterilization with additives did not compromise the mechanical properties of bovine cortical bone *[Russell et al., 2015]*.

The scCO_2 technology in combination with peracetic acid as additive has been commercialized for sterilization of various allograft tissues (Nova Sterilis, NY). A patent was issued in 1999 on scCO_2 sterilization, particularly polymers for drug delivery and implantation. The scCO_2 sterilization with additives is found to be less compromising on mechanical and rheological properties, as compared to low-temperature sterilization methods like gamma irradiation, ethylene oxide exposure, and steam sterilization. A confirmation test was carried out by comparing cytocompatibility of alginate gels and scaffolds with that of mineralized collagen after sterilization with ethylene oxide, gamma irradiation, steam sterilization, and scCO_2 treatment. It has also been reported that human mesenchymal stem cell viability and proliferation are not compromised by scCO_2 treatment *[Bernhardt et al., 2015]*.

Sterilization of polysaccharide hydrogels using scCO_2 with water, hydrogen peroxide, and acetic anhydride as additives is found effective in achieving SAL-6 for several bacterial species (post treatment up to 45min), except for *Bacillus pumilis* spores. Another study reports that dried alginate membranes can be sterilized using scCO_2 in the presence of hydrogen peroxide as sterilant. It is also observed that using the scCO_2 sterilization method it is possible to preserve the complex nanostructure of starch aerogel materials *[Ribeiro et al., 2019]*.

Thus, it is established that adding additives, such as, water, hydrogen peroxide, PAA, and acetic anhydride to scCO_2 is a very effective, non-cytotoxic, and promising alternative to the classical sterilization methods, especially for biomaterials.

9.7 Inactivation of Virus

Recently, disinfection of various viruses (*TGEV, PRV, JEV, and PRRSV*) spiked into a representative biomaterials (of different cell lines) was investigated by contacting with scCO_2 at 40–50°C, 160 bar for 45 min, and demonstrated that scCO_2 can effectively inactivate viruses in a heat-sensitive biomaterial without significantly decreasing bioactivity of the biomaterial *[Rao et al., 2016]*. The interaction of scCO_2 with virus envelop was attributed to be the predominant cause of inactivation. This is supported by the example of the lipid-coated *Sindbis* virus that is extremely resistant to high hydrostatic treatment (up to 7000 bar) *[Butz et al., 1992]*, while it is rapidly inactivated in presence of pure scCO_2. A commercial process was developed for preparation of bone allografts, based on pilot plant experiments (in a 2- liter autoclave), in which four viruses (HIV-1, *Sindbis,* Polio *Sabin* type I, and *Pseudorabies*) were loaded on bone samples and were contacted with scCO_2 for lipid extraction at 250 bar and 50°C for 10 min per g of bone. The virus load was heavily reduced and remained below detection level for any virus type (over 4 log reduction) *[Fages et al., 1997]*. Obviously, in this case, no care was taken for preservation of bio-molecules as only the mineral matrix was further used *[Perrut, 2012]*.

9.8 Sterilization Process for Rapid Inactivation of Bacterial Endospores

As noted earlier, $scCO_2$ treatment is a promising strategy for the terminal sterilization of sensitive biomaterials at low temperature. In combination with low amounts of additives $scCO_2$ treatment effectively inactivates microorganisms including bacterial spores. It has been reported that inactivation of *L.plantarum* could be achieved at 58 atm, 60°C and in 30-48 hours of treatment with $scCO_2$ *[Enomoto et al., 1997]*. Certain interesting research findings on biopolymers were established about the sterilization power of $scCO_2$ *[Dillow et al., 1999]*. Sterilization with $scCO_2$ at 30-60°C, 20.5MPa, and 0.6-4 hour of treatment time, could render complete inactivation of *S.aureus, B.cereus, L.innocua, S.salford, P.vulgaris, L.dunnifii, P.aeruginosa,* and *E.coli. Spilimbergo et al [2002]* gave a good insight into the mechanism by which $scCO_2$ inactivates microorganisms as discussed in detail in Chapter 8. It has been reported that spores could be inactivated by the said process at 8MPa and 75°C or 8MPa, 30 pressure cycles per hour at 35°C with 30min treatment time *[Spilimbergo et al., 2002]*.

Here a case study is reported on the development of a $scCO_2$-based sterilization process for achieving validated SAL10^{-6} levels of terminal sterilization at relatively low temperatures in a short time, with a minimum of moisture content. The process was investigated using a vessel constructed of stainless having a total internal volume of 600mL and also 20L *[White et al., 2006]*. The experimental set-up included heater strips and a method of internally stirring the fluid in an internal basket. The biological indicators used were *Bacillus atrophaeus* (*B. subtilis*) and *Bacillus stearothermophilus* spore strips (>10^6 CFU/strip) sealed in pouches and spore suspensions (>10^6 CFU/100μl in 40% ethanol). *Salmonella typhimurium* was cultured overnight in 500ml nutrient broth in shaker incubator at 37°C. *B. subtilis* and *B. stearothermophilus* spore strips after treatment, were cultured in 10ml nutrient broth at 37°C and 55°C respectively. It was demonstrated that it was possible to easily inactivate vegetative bacteria species including *E. coli* and *Salmonella typhimurium.* However, no log reduction, as compared to controls, was observed in 72 h, using dried commercially available spore preparations of *B. subtilis* or *B. stearothermophilus* endospores. In addition, inactivation of endospores did not appear to occur even with pressure cycling more than 30 times (200-100 bar) over 2 h, though the presence of water was found to facilitate inactivation of microbes with $scCO_2$ *[Dillow et al., 1999]*, as mentioned earlier.

Subsequently, an insightful study established the efficiency of $scCO_2$ as a terminal steriliser by using a combination of $scCO_2$ with low amounts of additives for the treatment and found that it effectively inactivates microorganisms including bacterial spores. *[Bernhardt et al., 2015]*. Several low molecular-weight volatile additives were assayed for screening of the effectiveness of additives and to identify the additives that could inactivate *B. stearothermophilus* endospores (*BI*). Out of these, as presented in *Table 9.3,* only trifluoroacetic acid (TFA) and peracetic acid (PAA) resulted in significant log reduction of *BI* at 100 bar pressure and respective temperatures. These two forms of acetic acid are characterized by relatively high vapor pressures, though PAA is non-toxic and unstable, whereas, TFA is exceptionally stable. PAA has been increasingly used in medical disinfection processes and it easily decomposes into acetic acid and water. Aqueous PAA is known to spontaneously reach a chemical equilibrium with acetic acid (AA) and hydrogen peroxide (HP).

Table9.3: Enhancement of *B. stearothermophilus* spore inactivation in $scCO_2$ with additives *[White et. al, 2006]*

Additive	Temperature (°C)	Time (h)	Log reduction
Ethanol	60-50	3	1.2-4.0
50% Citric Acid	60	2	0.03-0.62
Succinic Acid	50	2	0.25-0.29
Phosphoric acid	50	2	0.18-0.25
50% H_2O_2	50	1	0.13-1.57
Formic Acid	50	2	0
Acetic Acid	50	2	0.12-0.85
Malonic Acid	50	2	0-0.12
TFA	60	1	>6.4
5% PAA	60	1	>6.4

As demonstrated in *Table 9.3* the $scCO_2$ treatment at 100 bar and 60°C in combination with low concentration (5%) of the additive PAA is effective at inactivating bacterial endospores by log reductions of > 6.4 within 1h of treatment. The primary active component is PAA which in combination with $scCO_2$ is proven to be able to rapidly inactivate the spores, since HP or AA alone shows little activity as sterilizing agent in the $scCO_2$ process. Though PAA is the driving force for inactivation and the greatest log reductions in CFUs are consistently realized with higher PAA concentrations, HP appears to be responsible for eliminating some sporicidal activity.

The $scCO_2$ sterilization treatment with addition of 0.25% water, 0.15% hydrogen peroxide, and 0.5% acetic anhydride has been successfully validated for the inactivation of a wide panel of microorganisms including endospores of different bacterial species, vegetative cells of gram positive and gram negative bacteria including mycobacteria, fungi including yeast, and bacteriophages *[Bernhardt et al., 2015]*.

In order to ascertain whether the inactivation is observed when PAA is combined with $scCO_2$, is solely as a result of the PAA activity and has little to do with $scCO_2$ itself, experiments were conducted with the same concentration of PAA in conjunction with pressurized air using the same vessel and procedure. However, this level of inactivation was not observed with pressurized air. It is thus concluded that PAA along with $scCO_2$ proved to be nearly 100 times more effective than PAA with pressurized air and 5 times more effective than HP with $scCO_2$. This result would seem to suggest significant synergism between PAA and $scCO_2$ for the inactivation of endospores. Inactivation follows linear kinetics, making it possible to estimate SAL10^{-6} levels of sterilization by defining the *D* value (i.e., the time required for 1 log reduction) for the process.

Moreover, the process appears to be gentle, as morphology and protein profiles of inactivated bacteria are largely unchanged. PAA is both an acid and peroxide. As an acid, PAA may have unique transport properties in $scCO_2$ which also contribute to overall intracellular acidification. The same mass transfer enhancement may also facilitate the delivery and/or action of PAA as a sporicidal agent. This hypothesis is consistent with the synergy observed between $scCO_2$ and PAA for inactivating bacterial endospores. It has been also demonstrated that the bacterial spores are inactivated and not deactivated. These observations are all in the context of a terminal sterilization process that utilizes gas-permeable packaging, for a low temperature (≤40°C) and relatively low pressure (133 bar) process.

9.9 Elimination of Endotoxins and Pyrogens

For prostheses implantation and parenteral drug administration, it is imperative that the devices are sterile, but should be free of endotoxins and pyrogens that induce undesired side-effects like fever and inflammation. As these molecules are hydrophilic lipo-polysaccharides mainly originated from cell wall of gram negative bacteria, their elimination is very difficult. When removed by contacting with chemicals, the treated substrate is often damaged. Treatment with pure $scCO_2$ at 276 bar, 25°C for 120 min does not remove endotoxin. However treatment with $scCO_2$ with surfactant and water at the same condition successfully eliminates, the endotoxins completely, probably as a result of forming a water-in-CO_2 micro-emulsion that "washes" the endotoxins *[Perrut, 2012]*.

9.10 Sterilization of Biologically Active Large Molecules and Implant Materials

Many rapidly developing medical devices require novel sterilization solutions that involve the use of biologically active large molecules, such as, DNA, proteins, and bio-polymers. Of these, the increase in the use of human allograft tissue would be most benefitted by the $scCO_2$–based sterilization technology. This process for achieving SAL10^{-6} terminal sterilization may satisfy the need in tissue banking practice and other prospective commercially viable applications of terminal sterilization with $scCO_2$ including tissue, biodegradable polymers, powdered drug formulations, endoscopes, composite medical devices, DNA-based pharmaceuticals, and large molecule-based pharmaceuticals.

Different tests are performed to evaluate stability of biomaterials in terms of changes in the morphology, structure or mechanical properties of biomaterials after effective sterilization. It has been reported that the biomechanical properties of allografts sterilized with $scCO_2$ are superior to those sterilized by traditional gamma irradiation. The preservation of these biomechanical properties is important due to the fact that bone grafts are largely used in load bearing orthopedic applications. An increased anabolic response in their allograft bone chips is also reported when compared to sterilization by gamma irradiation, which may facilitate earlier healing clinically, as it did not affect allograft strength *[Russel et al., 2013]*. However, the $scCO_2$ sterilization resulted in significantly lower stiffness than unprocessed and irradiated allografts. The biological properties do not seem to have been affected after $scCO_2$ sterilization in these biological materials. It has been demonstrated that it is possible to sterilize catheters by $scCO_2$ without enhancing the binding capacity of microorganisms on the surface of the catheter shell, whereas ETO sterilization may induce a modification on the surface of the catheter shell, causing an increase in the binding capacity of microbial cells to the catheter shell *[Soares et al., 2019]*.

Several biological studies performed on natural biomaterials demonstrated good cytocompatibility and biocompatibility after sterilization with $scCO_2$. The allograft bone chips are biocompatible; the human amniotic membrane tissue is an excellent scaffold for the stem cell attachment and proliferation *[Jimenez et al., 2008]*.

Cytocompatibility of alginate gels and scaffolds from mineralized collagen was compared after sterilization with ethylene oxide, gamma irradiation, steam sterilization, and $scCO_2$ treatment. In addition, the aspects of viability and proliferation of human mesenchymal stem cells were evaluated. It was found that these were not affected and no adverse effects were established, as a result of $scCO_2$ treatment of these materials and scaffolds *[Bernhardt et al., 2015]*.

Further, the impact of $scCO_2$ sterilization was also tested by analyzing the mechanical properties of polysaccharide-based hydrogels and collagen-based scaffolds before and after the treatment in comparison to other processes. The physicochemical properties of polysaccharide- and collagen-based biomaterials were less compromised by $scCO_2$ compared to the classical and established low-temperature sterilization methods, like gamma irradiation and ethylene oxide exposure, as well as conventional steam sterilization. It was observed that, in the case of methylcellulose, the rheological parameters were not affected by $scCO_2$ sterilization while viscosity of gamma irradiated samples was dramatically decreased. Further, *Karajanagi et al. [2011]* demonstrated that polyethylene glycol hydrogels preserved their properties without compromising their structure, pH, water content, and viscoelastic properties when they are effectively sterilized with $scCO_2$, in contrast to the conventional methods of sterilization, and were found biocompatible and non-toxic when implanted subcutaneously in ferrets. Similarly Poly (L-lactic acid) porous scaffolds treated with $scCO_2$ retained their biocompatibility and structure *[Soares et al., 2019]*.

The development of bio-resorbable implant materials is rapidly increasing. Sterilization of those materials is inevitable to assure the hygienic requirements for critical medical devices according to the medical device directive (MDD, 93/42/EG). In particular, biopolymer-type of biomaterials are often highly sensitive towards conventional sterilization processes, e.g., steam treatment, ethylene oxide exposure or gamma irradiation.

It is established that there are no significant changes in most of the materials tested with respect to control materials. Both $scCO_2$ sterilization and ethanol+PAA methods were found to be efficient to provide 100% sterility in decellularized heart valves. However, $scCO_2$ sterilization does not cause damage and cross-link, unlike the ethanol+PAA sterilization or gamma irradiation and H_2O_2. The $scCO_2$ sterilization does not seem to greatly affect the properties of the materials tested or at least it appears to be less aggressive than with the conventional chemical methods.

Furthermore, for robust testing of the sterilization effect with respect to the implant to be used subsequently, the sterilization of Process Challenge Devices (PCD) was performed with all microorganisms-embedded in alginate/

agarose cylinders. These PCD served as surrogate models for bioresorbable 3D scaffolds. The PCD was subjected to $scCO_2$ treatment of scaffold precursor powders and scaffolds for 5, 15, 30, and 45 min, at ~8.5 MPa and ~38°C. All samples were treated in the presence of 0.25% water/0.15% H_2O_2/0.5% acetic anhydride. Depressurization was conducted in all cases within 6.5 min *[Bernhardt et al., 2015]*.

It has been thus established that the sterilization using $scCO_2$ with addition of 0.25% water, 0.15% hydrogen peroxide, and 0.5% acetic anhydride successfully inactivated a broad panel of microorganisms including bacterial endospores, even when embedded in hydrogel PCD and sealed in tyvek pouches. It is thus concluded that the sterilization process using $scCO_2$ with additives of water, hydrogen peroxide, and acetic anhydride is very effective, gentle, non-cytotoxic, and thus a promising alternative sterilization method especially for biomaterials.

References:

- Bernhardt, Anne; Wehrl, Markus; Paul, Birgit; Hochmuth, Thomas; Schumacher, Matthias; Schütz, Kathleen; Gelinsky, Michael; *"Improved Sterilization of Sensitive Biomaterials with Supercritical Carbon Dioxide at Low Temperature,"* PLOS ONE, June 12, 2015, DOI:10.1371/journal.pone.0129205.
- Butz, P., Habison, G.; Ludwig, H.; *"Influence of high pressure on a lipid coated virus"*, in: C. Balny et al (Ed.), *"High Pressure and Biotechnology"*, 224, John Libbey Eurotext Ltd., pp. 61–64., 1992.
- Dillow Angela K., Dehghani Fariba., Hrkach Jeffrey., Foster S., Neil R.; Langer Robert; *"Bacterial inactivation by using near and supercritical carbon dioxide,"* PNAS, 96 (18),10344-10348, August 31, 1999.
- Enomoto A, Nakamura K, Nagai K, Hashimoto T., Hakoda M., *"Inactivation of food microorganisms by the high-pressure carbon dioxide treatment with or without explosive decompression,"* Biosci Biotechnol Biochem 61:1133-1137, 1997.
- Fages, J.; Mathon, D.; Poirier, B.; Autefage, A ; Larzul, D. ; Jean, E.; Frayssinet, P.; *"Supercritical processing enhances viral safety and functionality of bone allografts"* in: S. Saito, K. Arai (Eds.), Proceedings 4th International Symposium, Supercritical Fluids, pp. 383–386, 1997.
- Guggenbichler, Josef Peter; Assadian, Ojan; Boeswald, Michael; Kramer, Axel; *"Incidence and clinical implication of nosocomial infections associated with implantable biomaterials - catheters, ventilator-associated pneumonia, urinary tract infections,"* GMS Krankenhhyg Interdiszip. 6(1): 2011; Doc18. doi: 10.3205/dgkh000175. Epub 2011 Dec 15. PMID: 22242099; PMCID: PMC3252661.
- Jimenez, A.; Zhang, J.; Matthews, M.A.; *"Evaluation of CO_2 based cold sterilization of a model hydrogel"*, Biotechnology & Bioengineering 101

1344–1352, (2008).

- Karajanagi, S.S.; Yoganathan R, Mammucari R, Park H, Cox J. Zeitels SM, Langer R, Foster NR, *"Application of a dense gas technique for sterilizing soft biomaterials"*, Biotechnol. Bioeng. 108 (7), 1716–1725, 2011.
- Lambert Byron J., Mendelson Todd A., Craven Michael D.; *"Radiation and Ethylene Oxide Terminal Sterilization Experiences with Drug Eluting Stent Products,"* AAPS Pharm SciTech.; 12(4): 1116–1126, Dec. 2011.
- Moore, M.B.; McCulley, J.P.; Luckenback, M., *"Acanthamoeba keratitis associated with soft contact lens,"* Am J Ophthalmol; 100: 396, 1985.
- Nichols, Anastasia *"Studies on the Sterilization of Human Bone and Tendon Musculoskeletal Allograft Tissue Using Supercritical Carbon Dioxide,"* J.Orthopaedics; 6(2) e9, 2009.
- Qiu Q.Q., Sun W.Q., and Connor J., *"Sterilization of Biomaterials of Synthetic and Biological Origin,"* Comprehensive Biomaterials vol. 4, pp. 127-144, (2011).
- Ribeiro, Nilza; Soares, Gonçalo C.; Santos-Rosales, Víctor; Concheiro, Angel; Alvarez-Lorenzo, Carmen; García-González, Carlos A.; Oliveira, Ana L.; *"A new era for sterilization based on supercritical CO_2 technology,"* J Biomed Mater Res. 1–30, 2019.
- Rao, L., Wang, Y.; Chen, F.; Liao, X.; *"The synergistic effect of high pressure CO_2 and nisin on inactivation of Bacillus subtilis spores in aqueous solutions",* Front. Microbiol. 7 1507, (2016).
- Rimondini, L.; Fini, M.; Giardino, R., *"The microbial infection of biomaterials: A challenge for clinicians and researchers. A short review,"* Journal of Applied Biomaterials & Biomechanics; Vol. 3 no. 1: 1-10, 2005.
- Russell, Nicholas; Rives, Alain; Matthew H. Pelletier, Wang, Tian; Walsh, William R.; *"The effect of supercritical carbon dioxide sterilization on the anisotropy of bovine cortical bone,"* Cell Tissue Bank,16:109–121, (2015).
- Russell, N.; Oliver, R.A.; Walsh, W.R.; *"The effect of sterilization methods on the osteoconductivity of allograft bone in a critical-sized bilateral tibial defect model in rabbits",* Biomaterials 34 (33) 8185–8194, (2013).
- Soares, Goncalo C.; Learmonth, David A.; Vallejo, Marian C.; Perez Davila, Sara; Gonzalez, Pio; Sousa, Rui A.; Oliveira, Ana L.; *"Review: Supercritical CO_2 technology: The next standard sterilization technique?",* Materials Science & Engineering C 99 pp. 520–540, (2019).
- Spilimbergo, S.; Elvassore, N.; Bertucco, A. *"Microbial inactivation by high-pressure",* J. Supercrit. Fluids, 22 pp. 55-63, (2002).
- Tessarolo, Francesco; Nollo, Giandomenico; *"Sterilization of Biomedical Materials",* Encyclopedia of Biomaterials and Biomedical Engineering,2501-2511, 2008.
- White, Angela,; Burns, David, Christensen, Tim W; *"Effective terminal sterilization using supercritical carbon dioxide,"* J. of Biotechnology, Vol 123, Issue 4, 10 June, pp. 504-515, 2006.

❐

10

Clinical Sold Waste Management with $ScCO_2$ Sterilization Technology

10.1 Generation, Handling, and Safe Disposal of Clinical Solid Wastes

In recent years, all over the world the varieties and magnitude of solid wastes generated from healthcare service providers (i.e., hospitals, nursing homes, pathological laboratories, and other supported healthcare services) have created major concern. In 1994, the World Health Organization (WHO) classified these wastes as *Health Care Wastes* (HCW) which are generated from the treatment, diagnosis, or immunization of humans and/or animals at hospitals, veterinary and health-related research facilities, and medical laboratories. This type of wastes contains infectious materials, toxic chemicals, and heavy metals, and may even contain radioactive substances. Generally, a fraction (10–25%) of the total HCW carries the infectious pathogenic microorganisms, hence all wastes do not fall in the category of clinical waste. Accordingly, HCW is conventionally classified as *general HCW* or non-clinical waste and *special HCW* or clinical healthcare waste. Further, the non-clinical wastes (e.g., packaging materials, such as, cardboard, office paper, left-over food, cans, etc.) are categorized under those which do not pose any hazards to health or environment, whereas, the Clinical *solid wastes* (CSW) are categorized under those which are controlled by the 1992-Waste Regulation Acts *[HMSO, 1992]*. WHO (in the year 2000) defined CSW as "any solid waste generated in the diagnosis, treatment, or immunization of human beings or animal, in

M. Mukhopadhayay and A. Chatterjee, *Sterilization and Preservation*,
https://doi.org/10.1007/978-3-031-17370-7_10

research pertaining to or testing of biological samples, including but not limited to soiled or blood-soaked bandages, culture dishes, and other glassware". CSW also includes discarded surgical gloves and instruments, needles, culture, stocks, and swabs used to inoculate cultures and remove body organs. CSW may be classified under different types of wastes, such as, infectious waste (e.g., those generated from the healthcare of patients suffering from infectious diseases), radioactive waste (e.g., radioactive substances including used liquids from radiotherapy or lab work), chemical waste (e.g., solid, liquid and gaseous chemicals from diagnostic and experimental work, cleaning materials), pathological waste (e.g., human tissues or fluids, body parts, blood, and other body fluids), and pharmaceutical waste (e.g., unused or expired drugs), and sharp objects (like sharps needles, syringes, blades, broken glass, scalpels, etc.) *[Hussein et al., 2011]*.

Recently COVID-19 pandemic and the subsequent governmental policies that were put in place to contain the spread of the virus, have created a global economic recession along with generation of a huge amount of wastes, especially medical wastes. This increase in the medical wastes specifically constituted of disposable plastic-based personal protective equipment (PPE) and single-use plastics by online shopping for most of the basic necessities, thus, in turn, led to altering the average density of the medical wastes.

The microbial agents that are present in medical wastes may contribute to the transmission and acquisition of infectious diseases during their handling and transportation. The relationship between the infectious agent, the host and the mechanism through which the microorganism gains access is referred to as the chain of infection, as explained in *Figure 10.1 [Salkin, 2004]*.

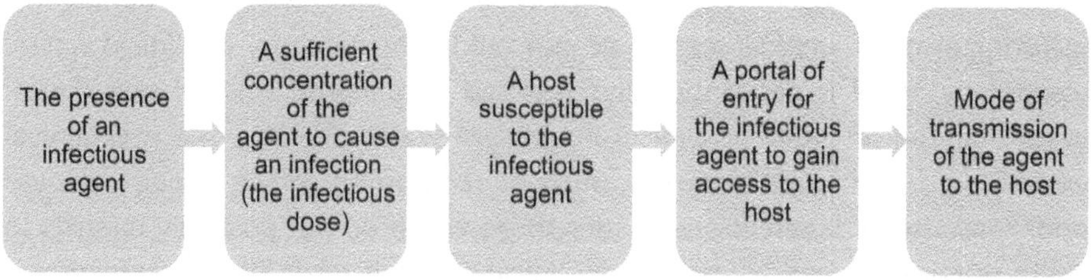

Fig. 10.1: Chain of infection during handling and transportation of medical wastes

In recent years, the treatment of clinical solid waste (CSW) is finding increasing importance, due to the infectious nature of the waste. Safe management of CSW backed by sustainability is not practised adequately, as there is a lack of proper regulation in most countries *[Singh et al., 2020]*. The management of CSW encompasses all aspects of supervision and control from its source to disposal, namely, generation to disinfection, collection to transportation, disposal to reuse, controlling of hazardousness to its ease of handling, and the economy of the entire operation. In general, emphasis is given to steps related to safe handling and disposal of the clinical waste.

Due to the presence of pathogenic bacteria, CSW may spread diseases and is potentially dangerous to the health workers during its handling, transportation, and safe disposal. Disposal of healthcare waste without following safety norms will result in polluting the environment and will also be conducive to the spread of infectious diseases, such as, Hepatitis, HIV/AIDS, cholera, typhoid, and respiratory complications, which are mainly caused by the reusing of the disposable medical equipment or by scavenging the medical waste *[WHO, 2018]*. Its impact on safety of health workers and environmental pollution entails a major challenge. In general, emphasis is given to better management and control of safe handling and disposal of CSW by developing specialized skills of health workers, awareness drive, and effective supervision and networking. Most importantly, with a view to reducing the financial burden, the healthcare providers have been constantly looking for cost effective programs of handling, disposal, and recycle-reuse of CSW after sterilization.

10.2 Recycle-Reuse of CSW after Sterilization

In general, the amount and rate of generation of clinical wastes are decided on the basis of the economic factors, such as, the costs of medical appliances and packaging substances, and the ease of the operation of reuse-recycling programs. Accordingly, the waste management plan needs to pay proper attention to the right segregation methods at the source of its generation so as to avoid contamination of non-clinical waste by clinical waste, thereby to reduce the proportion of the amount of clinical waste to the total waste generated.

An effective clinical waste management practice is governed by legislation and guidelines during waste handling, storage, and transportation to minimize the clinical waste generation. Also, an effective and economic clinical waste disposal program for the inactivation of possible pathogenic micro-organisms present in CSW must satisfy the following criteria: (i) minimal risk assessments, (ii) minimal health impacts, (iii) minimal environmental impacts, and (iv) cost effectiveness.

The strategy of recycling-reuse after sterilization makes the process of management of CSW economically rewarding. It not only renders value addition to CSW, but also reduces the risks to healthcare facilities. In addition, if the sterilization is carried out at the point of initial collection, the handling and transportation problems are eased out. In that case, CSW would, not only pose any threat to healthcare workers, but also render economic benefits. The recycling-reuse strategy may be carried out even with the non-skilled health workers, decreasing the labor costs. Thus, healthcare enterprises can economically provide a safe environment for patients, healthcare workers, and derive better compliance to regulatory obligations.

The continuously increasing costs for treatment and disposal of CSW are prompting to explore alternative management routes. It is reported, that disposal of general healthcare wastes (HCW) may require 10-20 times lower costs than that of CSW. Also, HCW may contain large volumes of materials which can be reused and recycled after sterilization. In other words, only a fraction of HCW is required to be sterilized for the development of a cost-effective management of recycling-reuse program of CSW. Thus, the recycling and reuse of CSW reduces the volume of waste generation as well as the disposal cost.

Currently, in some developed countries, the recycle-reuse programs are in operations in hospitals for waste materials, like office paper, cardboard metal cans, and selected glass. For this purpose, waste materials are needed to be segregated and classified at the place of their generation, in order to protect the collecting staffs from the infectious nature of CSW and develop an effective technology to sterilize CSW at the source of generation. The use of supercritical carbon dioxide ($scCO_2$) technology addresses most of the above concerns. Though the priority of this sterilization technology is to inactivate the pathogenic micro-organisms present in CSW, it would also enable the recycling-reuse of CSW materials after sterilization.

10.3 Pathogenic Microorganisms Present in Clinical Wastes

Clinical wastes are highly infectious owing to a large number of pathogens, like, viruses, microbes, fungi, spores, and bacteria. It is mandatory to sterilize them before they are discharged into the environment. For assessing the health risk of the clinical wastes, it is necessary to identify the various types of pathogens, bacteria and fungi that have the ability to multiply and persist for a long time in clinical wastes. Microbiological analysis of CSW conducted to determine the quantity of infectious microorganism by colony count methods revealed the presence of a wide variety of bacteria in CSW *[Efaq et al., 2015]*, as listed in *Table 10.1*. As viruses cannot survive in clinical wastes without a host organism, it is important to know the environmental factors of clinical wastes, such as, presence of carbon and nitrogen sources (e.g., sugars, protein, starch, fats, and other compounds), and factors that are suitable for growth and survival of bacteria and fungi.

Table 10.1: Pathogenic microorganisms present in CSW *[Efaq et al., 2015]*

Group	Indicator Microorganism
Vegetative Bacteria	*Staphylococcus aureus* (ATCC 6538) *Pseudomonas aeruginosa* (ATCC 15442)
Fungi	*Candida albicans* (ATCC 18804) *Penicillium chrysogenum* (ATCC 24791) *Aspergillus niger*
Viruses	Polio 2 or Polio 3 MS-2 Bacteriophage (ATCC-15597-81)
Mycobacteria	*Mycobacterrium terrae* *Mycobacterrium phlei* *Mycobacterrium bovis* (BCG) (ATCC 35743)
Bacterial spores	*Bacillus stearathermaphilus* (ATCC 7953) *Bacillus subtilis* (ATCC 19659)

10.3.1 Pathogenic Bacteria in Clinical Wastes

Clinical wastes may contain a wide range of pathogenic bacteria; the types of these pathogens depend on the source and composition of clinical wastes. Pathogenic bacteria in the diagnostic specimens, lab cultures, and other healthcare facilities are likely contaminants in clinical wastes. *Park et al [2009]* established the presence *Pseudomonas spp., Lactobacillus spp., Staphylococcus spp., Micrococcus spp., Kocuria spp., Brevibacillus spp., Microbacterium oxydans, Propionibacterium acnes* spp., *Staphylococcus aureus Coliform* bacteria, *Escherichia coli, Enterobacter spp., Pseudomonas aeruginosa, Bacillus cereus, Salmonella spp., Klebsiell spp., Enterobacter spp., Corynebacterium diphtheria, Legionella spp.*, yeast and molds in CSW collected from five major hospitals *[Efaq et al., 2015]*.

It is reported that *Hossain [2013]* isolated *Acinetobacter lwaffii, A. baumannii, E. coli, E. faecalis, K. pneumonia, Proteus mirabilis, Serratia marcescens, S. liquefaciens, Salmonella spp., S. aureus, P. aeruginosa, S. epidermidis, S. pyogens, and S. agalactiae* from CSW collected from different sections of hospital (e.g., dental section, microbiological laboratories, dermatology, obstetrical and gynecology units). *E. coli, E. faecalis, K. pneumonia, P. mirabilis, P. aeruginosa, S. aureus, S. pyogens, and Salmonella spp.* are the pathogenic bacteria in CSW, which are known to develop resistance to antibiotics upon the disposal into the environment.

It is known that pathogenic bacteria have the potential to grow and multiply in clinical wastes after 24h of storage period to reach the infective dose, and accordingly CSW is needed to be sterilized within this time. *Salmonella spp.* is the most significant pathogen, which can cause diseases to all organisms from insects to mammals. Enteric fever is a collective term given to the invasive infections caused by *S. typhimurium,* the cause of typhoid fever, and by the strains of *S. paratyphi* that cause paratyphoid fever. Enterohaemorrhagic strain *E. coli* O157:H7 can cause gastrointestinal disorders, such as, bloody diarrhea, cramping and abdominal pain. *E. coli* is accountable for neonatal meningitis and pneumonia infection. *P. mirabilis* has the ability to survive for long time in the healthcare facilities and hospitals. *K. pneumonia* causes bronchopneumonia and bronchitis [*CDC, 2009*]. The wide spectrum of pathogenic bacteria present in CSW suggests the health risk associated with the wastes which call for an efficient treatment processes to prevent the pathogens from entering into the environment and the humans.

10.3.2 Infectious Fungi in Clinical Wastes

Many infectious fungi may be generated from hospitals and other healthcare facilities or clinical laboratories and can cause several serious diseases to human. Of particular concern are the opportunistic pathogens e.g., *C. neoformans, Candida spp., Aspergillus spp., Mucorales, Fusarium spp., Paecilomyces spp., Alternaria spp*., and Scedosporium spp., which are also known to cause invasive fungal infections *[Efaq et al., 2015]*. Out of these, *Aspergillus spp.* is commonly encountered in hospital wastes and can cause many diseases in human. The spores of *Aspergillus spp.* have the ability to survive for a long time in the environment. *Penicillium spp.* is the most common fungi in the environment and can be dangerous to immune-compromised patients. Studies revealed presence of a wide range of fungi, e.g.,. *Aspergillus spp., Fusarium spp., a Mucor spp., and Paecilomyces spp*., on hospital fabrics and plastic objects. Fungi also have the ability to live in clinical wastes disposed into landfills. The reproduction of fungi in dumping fields may give rise to distribution of these spores in air. Therefore, even if the clinical wastes generated from different sections of healthcare facilities are free from fungi, these wastes may generate contaminants during storage before the final treatment. *Efaq et al [2015]* studied the fungal diversity in clinical wastes generated from a healthcare facility and reported that fungi detected in 83.7 % of the samples constituted of *Aspergillus Nigri, A. niger, A. fumigatus, A. tubingensis, P. simplicissium, P. waksmanii, and C. lunata*, most of which indicated the ability to infect the human. Thus, the fungi constitute a major source of contaminants in CSW due to the constant presence of necessary nutrients for fungal growth as well as the ability of fungal spores to survive in the environment.

10.4 Thermal methods for Inactivation and Neutralization

The efficiency of inactivation process is determined in terms of log reduction and inactivation rate of fungal spores. Any treatment technology for clinical wastes should have the ability to inactivate the biological indicators by 6 log reductions with initial concentration of 10^6 cells/mL, as per *STAATT [2005]*. Many treatment technologies have been employed to inactivate these infectious pathogens to sterilize CSW for safe disposal into the environment. The principles behind these technologies are based on the factors responsible for survival and growth of these microorganisms. Some of these factors include temperature, pressure, pH, moisture content, and environmental factors, such as, presence of nutrients, irradiation or disinfectants.

The main aim to treat or decontaminate medical waste, is to contain the spread of infectious disease and render it safe for handling and disposal. The most primitive disposal methods of CSW are open dumping, landfill or incineration. Other recent disposal methods include steam sterilization or autoclaving, chemical sterilization, microwaving, etc. Though microwaving CSW may be economically competitive compared to the incineration, but microwave technology may not be amenable to large scale of operations. It is also reported that microwaving of CSW may be inadequate for sterilization of all pathogenic microorganisms.

The various thermal treatment methods for this purpose are incineration, steam sterilization (i.e., autoclaving), microwave techniques, or interment (for anatomy wastes). The incinerators, autoclaves and microwaves techniques have been studied extensively for inactivation of pathogens in the clinical wastes *[Banana et al., 2013]*. Alternative treatment methods developed over the last few years include chemical disinfection, grinding/shredding/disinfection methods, energy-based technologies (e.g., microwave or radio wave treatments), and disinfection/encapsulation methods *[Centre for Disease Control and Prevention, 2003]*.

10.4.1 Incineration

Temperature is one of the critical factors affecting survival of microorganisms, and as a result this is a basic parameter on which many sterilization technologies have been employed. The most common technology applied for the inactivation of the pathogens is incineration. The incineration process uses high temperature between 900°C and 1200°C, which destroys microorganisms rapidly. The incineration is reported to be the most effective treatment to achieve the standard sterility levels. Application of incineration temperature as low as 300°C for 15 min or very high temperature of 1100°C for

3 min seems to be able to attain the required sterilization assurance level of 10^6 reduction *[Blenkharn, 2005]*. However, this process has adverse effects on the environment and humans.

10.4.2 Steam Sterilization

In some thermal methods the temperature may be combined with other factors, such as, extended time and moisture content to enhance the inactivation rate. For example, temperature between 160°C and 180°C for 2h is sufficient to destroy the microorganisms. In autoclaving, the moist temperature, pressure, and time are the prime factors for the neutralization of microorganisms. It is reported, that the sterilization process conducted at 121°C below 2 bar pressure for 15 min destroys the microorganisms by the irreversible denaturation of enzymes and structural proteins *[Lee et al., 2004]*.

It is well known that pathogenic bacteria require a certain temperature for growth in the range between 10°C and 40°C, with optimum temperature being 37°C, while that for fungi to grow well is 28°C. The fungal spores can be destroyed at 70°–80°C in10 min. These facts imply that the risk associated with the pathogenic microorganism is neutralized by increasing the temperature above this range during the thermal treatment process owing to the denaturation and inactivation of metabolic enzymes of the microorganism cells. For many years, steam autoclave has been used for sterilization of medical instruments in hospitals. However, in the clinical waste, high contents of microorganisms along with a more complex matrix affect the efficiency of steam autoclave treatment of clinical wastes.

For treating medical wastes in a gravity flow autoclave, the standard practice comprises: (i) treatment for 60min at 121°C and 1bar, or (ii) treatment for 45 min at 135°C and 2 bar, or (iii) treatment for 30 min at 149°C and 3.5 bar. The treatment should completely and uniformly destroy the biological indicator which usually is *Bacillus stearothermophilus* spores using vials or spore strips, with at least $1x10^4$ spores per milliliter.

The other thermal methods of sterilization for the safe management of CSW include heating in microwave. In microwaving the microorganisms are inactivated by the rise in temperature in addition to electromagnetic radiation *[Watanabe et al., 2000]*. Thermal treatment using microwave also has the ability to achieve the 6 log reduction required for the microorganisms.

10.5 Nonthermal Sterilization Technology Using Supercritical CO_2

The selection of the appropriate technology for the treatment of the clinical waste should consider the adverse effects on the environment, in addition to the ability to inactivate the pathogens. Moreover, the sterilization technology should be easily implementable, ecofriendly, with no toxic by-products and no chemical additives *[WHO, 2005]*. Two new concepts have emerged in the last two decades, namely, the Climate Change (Global Warming) and Zero Discharge. Global Warming implies the increase of the earth temperature due to emissions of gases (CO_2 and CH_4) into the atmosphere. It is indicated that by the end of this century, the global surface temperature may rise in the range 1 to 3.4°C due to the adverse effects of deforestation and industrialization. Zero discharge is a term frequently advocated in the context of minimizing environmental pollution by adopting recycling or reuse program for management of wastes, particularly for industrial waste water treatment.

Sterilization using $scCO_2$ is envisaged as a potential alternative technology using a non-thermal process of sterilization for management of CSW. The attractive feature of this process is that it can be operated at near ambient temperature under moderate pressure as in the case of food and pharmaceutical industries. Studies on non-thermal method of sterilization are first carried out on food, normal saline, and growth media to assess the effectiveness of this method on the inactivation of bacteria and fungi present in CSW. Subsequent studies on CSW indicate that $scCO_2$ sterilization is effective in inactivation of the pathogenic bacteria present in CSW.

10.5.1 Inactivation of Micro-Organisms Using $ScCO_2$

As mentioned earlier in Chapter 9, the $scCO_2$ sterilization has the potential to sterilize biomedical device for being effective against bacteria, viruses, and spores, though in the case of some spores, additives like H_2O_2 or PAA are needed to be used for terminal sterilization. However, *STAATT [2005]* recommended that the accurate evaluation of alternative technologies for inactivation of biological indicators in clinical wastes should be conducted without any chemical inactivation agent.

The efficiency of $scCO_2$ for inactivation of microorganisms is dependent on several factors, such as, temperature, pressure, time, moisture content, and chemical additives, though the physical and chemical characteristics of the suspending medium play an effective role on the inactivation of microorganisms. Among these factors, pressure and temperature influence the inactivation rate most *[Hossain, 2013]*. Moreover, temperature has more influence on the inactivation rate at low pressure and water activity and pressure showed insignificant influence at high temperature and water activity. The treatment time is reduced, as CO_2 can easily penetrate into the microorganisms and it does not need to be saturated in the medium to attain the required pH value for inactivation. The presence of moisture in $scCO_2$ is found to enhance the inactivation rate of microorganisms, due to external pH reduction and promotion of membrane cell swelling and rupture. $ScCO_2$ is more effective in the presence of water. A few case studies have been reported on sterilization with and without additives for different types of pathogens, like viruses, microbes, fungi, spores, and bacteria, etc., present in CSW as listed in *Table 10.2 [Hossain et al., 2011].*

Table 10.2: Micro-organisms inactivated by $scCO_2$ at different conditions *[Hossain et al., 2011]*

Micro-organisms	Experimental conditions				References
	Pressure bar	Temp °C	Time min	Additive	
Salmonella enterica	80-250	35-55	15-30		*Kim et al., [2009]*
Bacillus subtilis spores	70-150	36-75	120		*Spilimbergo et al., [2003]*
Baker's Yeast	203	35	120	Ethanol/ Acetic Acid	*Kamihira et al., [1987]*
Escherichia coli Staphylococcus oureus Conidia Lactobacillus brevis	250	35			*Ishikawa et al., [1995]*
Saccharomyces cerevisiae Geobacillus Stearothermophilus Bacillus atrophaeus Spores	304	40	60	H_2O_2	*Hemmer et al., [2006]*
Saccharomyces cerevisiae (Yeast)	70-210	>35	15		*Lin et al., [2008]*
Pseudomonas fluorescens	103-483	40	15-35		*Werner and Hotchkiss, [2006]*
Staphylo aureus & E. coli	40	60		H_2O_2	*Jimenez et al., [2008]*
Salmonella typhimurium	80-150	35-45	10-50		*Kim et al., [2007]*
Bacillus pumilus spores	275	60	240	H_2O_2	*Zhang et al., [2006 b]*
Bacillus atrophaeus spores	275	40	240	H_2O_2	*Zhang et al., [2006 c]*

Moreover, *Hossain [2013]* used glycerol as a surfactant for the homogeneous distribution of bacteria in clinical solid wastes for inactivation by $scCO_2$. It has been reported since 1940 that the glycerol reacts with CO_2 to produce succinic acid. *Wu et al [2011]* indicated that this acid has inhibited growth and conidia germination of *F.oxysporum*. Thus, inactivation of fungi using $scCO_2$ in the presence of glycerol might be due to succinic acid and not to the action of $scCO_2$.

10.5.2 Inactivation of Fungi Using $ScCO_2$

Efaq et al [2016] investigated the inactivation of microorganisms under different conditions of pressure, temperature, and processing time using $scCO_2$. Almost 100% inactivation of initial fungal spores was obtained with a 6log reduction, under very high pressure of 35 MPa and a high temperature of 75°C in 90 min. This study confirmed that without addition of any chemicals, the sc-CO_2 sterilization method could be potentially used for the inactivation of microorganisms in clinical waste.

The optimal condition for inactivation of bacteria may not be appropriate for fungi. The studies on inactivation of fungi using $scCO_2$ *[Haas et al., 1989]* reveals that *P. roqueforti* spores in the growth medium can be reduced by 5 log at 5.4 MPa (subcritical CO_2), 45°C in 120 min and by 6 log at 5.4 MPa , 45°C in 240 min. Similarly, it is reported *[Shimoda et al., 2002]* that *A. niger* can be inactivated by 3 log reduction at 19 MPa, 46°C in 1.7 min, by 5 log reduction at 19 MPa, 48°C in 1.8 min, and by 6.8 log at 19 MPa, 50°C in 1.5 min from physiological saline. *Calvo et al [2007]* investigated the inactivation of *A. niger* and *A. ochraceus* spores in cocoa by $scCO_2$. They recorded an inactivation of both fungal spores at 30 MPa, 80°C and after 30 min in 5 % of water. *Table 10.3* presents studies conducted on inactivation of fungi using $scCO_2$ with some additives *[Efaq et al., 2015]*.

Table 10.3: Inactivation of fungal spores by treatment with $scCO_2$ *[Efaq et al., 2015]*

Fungal Species	Sample Medium with % additive	Treatment condition	Log reduction	Reference
A. niger	90% water Ethanol or acetic acid	20 MPa, 35ºC, 60 min	6	*Kamihira et al., [1987]*
A. niger	Physiological saline	19 MPa, 50°C, 1.5 min.	6.8	*Shimoda et al., [2002]*
A. niger	Cocoa with 5% water	30 MPa, 80°C, 30 min.	5.3	*Calvo et al [2007]*
P. roqueforti	Growth medium	5.4 MPa, 45°C (subcritical CO_2), 240 min.	5 6	*Haas et al [1989]*
Byssochlamys fulva	Ringer solution	5.0 MPa, 80°C, (subcritical CO_2), 85.5 min.	1	*Ballestra and Cuq [1998]*
A. ochraceus	Cocoa with 5% water	30 MPa, 80°C, 30 min.	3	*Calvo et al [2007]*
A. brassicicola	16% Ethanol	15 MPa, 38°C, 45 min.	7	*Park et al [2012]*
P. oxalicum	Ethanol	10 MPa, 40°C, 45 min.	Complete inactivation	*Park and Kim [2013]*
A. ochraceus	Water	7 MPa, 50°C, 5.0 min.	Complete inactivation	*Neagu et al* [2014]
A. niger Penicillium simplicissimum	CSW	35 MPa, 75°C in 90 min.	Complete inactivation	*Noman et al [2016]*

In addition, several studies were conducted earlier *[Jimenez et al., 2008; Kamihira et al., 1987; Spilimbergo et al., 2002]*. From these studies it can be concluded, that sterilization using $scCO_2$ could be a cost effective, easily implementable, and environment-friendly process for management of CSW as it is able to kill a wide variation of micro-organisms and the sterilized product can be reused.

10.6 Sterilization-Reuse of Personal Protective Equipment (PPE) Using $scCO_2$ Technology

In the context of management of COVID-19 pandemic, the world Health Organization (WHO) has given directives to the Governments of all countries to utilize the personal protective equipment (PPE), face masks, and respiratory masks to reduce the risk to healthcare workers around the world, and also to protect healthy people in community against the highly infectious virus. These are now some of the most demanding items worldwide as their use is not limited to the medical fraternity but the entire population present on earth is dependent on these masks to shield them from COVID-19. Face masks are mostly the surgical masks worn to protect environment, and near people from the wearer's respiratory emissions, and also to provide the wearer protection against large droplets, splashes, or sprays of bodily or other hazardous fluids. The respiratory masks (such as, N95, FFP2, and equivalents), are worn to reduce wearer's exposure to particles including small particle aerosols and large droplets. On the other hand, the excessive use and consumption of single-use PPEs have become a severe threat to the entire environment due to their micro and nano-plastic pollution *[Aragaw, 2020; Fadare and Okoffo, 2020]*. Negative environmental impacts of COVID-19 are emerging as a consequence of increased generation of medical waste *[Caraka et al., 2020]* and the decline in waste recycling *[Zambrano-Monserrate et al., 2020]*. There is currently no unified regulation on the mask and their pollution management, but this $scCO_2$-based sterilization and reuse of PPE could minimize the environmental risk.

In order to alleviate the global problem of shortage of single-use personal protective equipment (PPE) as well as their contaminating the environment by their disposal, there have been vigorous research attempts by many universities and governmental organizations to deal with this unprecedented crisis of COVID-19 pandemic. To sustain the availability of PPE, the strategies advocated by WHO include increasing supply of PPE by effective sterilization of PPE in large numbers, reprocessing them, extending the use and/or limited reuse of these items, and releasing stockpiled devices, that have passed their shelf life. There is an urgent need for evolving the strategies for increasing the emergency supply of these PPE through the effective application of sterilization process and improving the efficacy in their reuse. $ScCO_2$ could be a sanitization and reuse platform for decontamination of surgical, cloth, and N95 masks. $ScCO_2$ has proven its specific advantages over other current sterilization processes towards safe and effective decontamination and reuse of the PPE including masks, gloves, apparels, etc.

Very recently, the Novaclean process of sterilization using $scCO_2$ has been evaluated for decontamination and reprocessing of personal protective equipment (PPE), surgical masks, cloth masks, and N95 respirators from human coronavirus, HCoVNL63, and SARS-CoV-2 in the Center for Applied NanoBioscience and Medicine, the University of Arizona. *[Bennet et al., 2021]*. The NovaClean process is a low temperature, minimally reactive, deep penetrating, bioburden reducing, and decontaminating $scCO_2$ technology.

For preliminary trials, *Bacillus atrophaeus* spores (ATCC 9372; Crosstex, USA) were selected as a surrogate for validation for coronavirus and were inoculated into different samples corresponding to dry, hydrated, and saliva medium to simulate coughing and sneezing methods of contamination through PPE. The NovaClean process using $scCO_2$ has been described in detail elsewhere *[Bernhardt et al., 2015]* and *[Qiu et al., 2009]*. In this process the contaminated samples are placed inside the Nova Genesis 500 ml vessel and mixed with 7.5 mL of 75% ethanol with and without 15% H_2O_2 or peracetic acid (PAA)-based NovaKill™ sterillant. The vessel is operated at 10.0 -13.3 MPa of $scCO_2$ pressure at 33°C–35°C for treatment time of 5 to 90 min. The inactivation of the microbes by $scCO_2$ sterilization with H_2O_2 sterilant (NovaKill agent) was investigated to analyze the effects of pressure, temperature, mixing velocity, and composition of the additives. Also, human coronavirus SARS-CoV-2 and HCoV-NL63 were inoculated on the respiratory material, and viral activity was determined post-treatment. After decontamination, all spore strips/membranes were retrieved from the system and analyzed for inactivation/death. Moreover, the reprocessing ability of $scCO_2$-based decontamination was evaluated by wettability testing and surface mapping. The viral sanitization results showed a complete inactivation of both coronavirus HCoV-NL63 and SARS-CoV-2. No changes were observed in PPE morphology, topographical structure, or material integrity, and in accordance with the WHO recommendation, wettability was maintained post-processing. This way the critical elements for the decontamination and reuse of PPE were evaluated in any setting and thus this state of the art, current research provides a direction for implementation of the $scCO_2$ technology towards solving future availability of PPE and environmental protection.

This recent research demonstrated that the NovaClean $scCO_2$-based decontamination process has the ability to (i) inactivate *Bacillus atrophaeus* endospores in a 40-minute cycle irrespective of the environment of the spore (i.e., in dry/hydrated, coated in saliva medium), (ii) inactivate the most infectious human coronavirus strains, including SARS-CoV-2, being inoculated on respirator material, and (iii) maintain surface integrity and wettability of surgical, N95 and cloth masks. The low operating temperature, efficient and deep penetrating capability of $scCO_2$ and its lethality, when combined with a minimal amount of an oxidative additive H_2O_2 and peracetic acid in 75% ethanol can make the $scCO_2$-based decontamination an ideal solution for the healthcare setting. This technology offers the most desirable solution during the periods of pandemic.

10.7 Environment and Health Protection Using $scCO_2$ Sterilization Technology

The $scCO_2$ sterilization technology can be effectively utilized to sterilize the clinical solid waste at the point of initial collection, in order that collection, segregation, transportation, and recycling-reuse activities can be carried out at the point of generation. This way, healthcare centers can provide a safe environment, minimizing the chances of infection to the patients, people and health workers without contamination of the environment. This will decrease labor work force, minimize management costs, and yet yield better compliance to regulations. Therefore, the clinical solid waste management using $scCO_2$ technology may be implemented by adopting the following sequencing:

(i) segregation of recyclable, reusable and disposable items,

(ii) cleaning and disinfection of the pre-sterilized single-use items before disposal,

(iii) $scCO_2$ sterilization of reusable items, monitoring sterility assurance level,

(iv) recycling and reuse of the sterilized plastic materials.

Table 10.4 gives a brief insight of the various exiting methods and alternatives that are available, and can be followed for safe CSW management and disposal *[United Nations Environment Program, 2020]*. It is thus recommended to implement the $scCO_2$ sterilization technology with some additives at moderate pressure and temperature conditions. There is a serious need for efficient and safe management of clinical solid waste by the 'sterilize and reuse' program at the point of initial collection or by neutralization prior to final disposal in order to prevent infection and contamination.

Table 10.4: Common methods and alternatives for safe management of CSW.

Clinical Waste Management	Common Methods	Available Alternatives
Source separation	• Separation into infectious and non infectious , dry and wet	• Labelling and packing the waste (Infectious/Non-infectious) • Using double layered bags • Keeping track of the waste • Disinfection of bags before being dispatched
Storage	• Specially designated rooms	• Use of cold room • Disinfection of storage area on regular basis • Separation of infectious waste from other HCW in the storage room
Transport	• Use of PPE for transportation workers	• Use of specific vehicles and equipment for transportation of waste • Timely and frequent collection • Disinfection of bags/bins prior to loading the vehicle
Treatment	• Use of incineration • Use of specific landfill sites • Use of autoclaves	• Plasma pyrolysis • Dry heat • Cement Kiln • Deep burial • Open Landfill • Use of $scCO_2$ sterilization for reuse and recycle

References:

- Aragaw, T.A., *"Surgical face masks as a potential source for microplastic pollution in the COVID-19 scenario"* Mar. Pollut. Bull. 159, 111-517, 2020.
- Ballestra, P, Cuq, J.L., *"Influence of pressurized carbon dioxide on the thermal inactivation of bacterial and fungal spores"*, LWT Food Sci Technol 31:84–88, (1998).
- Banana, A.A.S., Norulaini, Nik, Baharom, J.N.A., Zuraida, M.Y.L., Rafatullah M, Kadir, Ab M.O., *"Inactivation of pathogenic micro-organisms in hospital waste using a microwave",* J. Mat Cycles Waste Manag 15(3):393, (2013).
- Bennet Devasier; Harris Ashlee F.; Lacombe, Jerome; Brooks, Carla; Bionda, Nina, Strickland, Aaron D.; Eisenhut, Tony; Zenhausern, Frederic; *"Evaluation of Supercritical CO_2 sterilization efficacy for sanitizing personal protective equipment from the coronavirus SARS-CoV-2"* Science of The Total Environment Volume 780, 1 August 2021, 146519.
- Bernhardt, A., Wehrl, M., Paul, B., Hochmuth, T., Schumacher, M., Schutz, K.; *"Improved sterilization of sensitive biomaterials with supercritical carbon dioxide at low temperature",* PLoS One 10, e0129205, (2015).
- Blenkharn, J. I., *"Safe disposal and effective destruction of clinical wastes",* J Hosp Infec 60:295–297, (2005).
- Calvo L, Muguerza B, Cienfuegos-Jovellanos E, *"Microbial inactivation and butter extraction in a cocoa derivative using high pressure CO_2"*, J Supercritical Fluids, 42: 80–87, (2007).
- Caraka, R.E., Lee, Y., Kurniawan, R., Herliansyah, R., Kaban, P.A., Nasution, B.I., et al., *Impact of COVID-19 large scale restriction on environment and economy in Indonesia,* Glob. J. Environ. Sci. Manag. 6, 65–84, 2020.
- Centre for Disease Control and Prevention, (CDC) *"Background I. Regulated Medical Waste Guidelines for Environmental Infection Control in Health-Care Facilities"* (2003).
- Fadare, O.O., Okoffo, E.D; *"Covid-19 face masks: a potential source of microplastic fibers in the environment",* Sci. Total Environ. 737, 140279, 2020.
- Efaq, N., Rahman, N.N.A., Nagao, H., Al-Gheethi, A.A, Shahadat, M., and. Kadir, M.O.A, *"Supercritical Carbon Dioxide as Non-Thermal Alternative Technology for Safe Handling of Clinical Wastes"* Environ. Process 2:797–822, (2015).
- Efaq A. Noman, Rahman, Nik N. N. A.; Shahadat, Mohammad; Nagao, Hideyuki; Al-Karkhi, Abbas F.M; Al-Gheethi, Adel; N.T. Lah. Tengku, Omar, Abdul K.M.; *"Supercritical Fluid CO_2 Technique for Destruction of Pathogenic Fungal Spores in Solid Clinical Wastes,"* Clean Soil, Air, Water, Vol 44, Issue 12, pp. 1700-1708, (2016).
- Haas, G.J., Prescott, H.E., Dudley, E., Dik, R., Hintlian, C., Keane, L. *"Inactivation of microorganisms by carbon dioxide under pressure",* J. Food Safety 9:253–265, (1989).

- Hossain S., *"Supercritical carbon dioxide sterilization of clinical solid waste"*, Ph.D Thesis, Environmental Technology Division, School of Industrial Technology, University Science Malaysia. Penang, Malaysia. (2013).
- Hossain, Md. Sohrab, Amutha, N.A., Norulaini, Nik, Omar, A.K. Mohd., *"Clinical solid waste management practices and its impact on human health and environment –A review"*, Waste Management 31, pp. 754–766, (2011).
- Jimenez, A., Zhang, J., Matthews, M.A., *"Evaluation of CO_2 based cold sterilization of a model hydrogel"*, Biotechnol Bioeng 101:1344–1352, (2008).
- Kamihira M., Taniguchi M., Kobayashi T., *"Sterilization of microorganisms with supercritical and liquid carbon dioxide"*, Agricul Biol Chem 51:407–412, (1987).
- Lee, B.K., Ellenbecker, M.J., Moure-Eraso, R., *"Alternatives for treatment and disposal cost reduction of regulated medical waste"*, Waste Manag 24:143–151, (2004).
- Neagu, C., Borda, D., and Erkmen, O., *"Mathematical modelling of Aspergillus ochraceus inactivation with supercritical carbon dioxide – A kinetic study"*, Food Bioprod Proces, 92(4):369–375,(2014).
- Park, H.S, and Kim, K.H., *"Enhancement of supercritical CO_2 inactivation of spores of Penicillium oxalicum by ethanol cosolvent"*, J. Microbiol Biotechnol 23(6):833–836, (2013).
- Park, H.S., Choi, H.J., Kim, K.H., *"Inactivation of Alternaria brassicicola spores by supercritical carbon dioxide with ethanol entrainer"*, J. Microbiol Methods, 88:185–187, (2012).
- Qiu, Q.Q., Leamy, P., Brittingham, J., Pomerleau, J., Kabaria, N., Connor, J., *"Inactivation of bacterial spores and viruses in biological material using supercritical carbon dioxide with sterilant"*, J. Biomed Mater Res B. Appl. Biomater 91, pp. 572–578. 2009.
- Salkin, Ira F.; *"Review of Health Impacts from Microbiological Hazards in Health-Care Wastes"*, 2004, WHO.
- Shimoda, M., Kago, H., Kojima, N., Miyake, M, Osajima, Y., Hayakawa, I., *"Accelerated death kinetics of Aspergillus niger spores under high-pressure carbonation"*. Appl. Environ. Microbiol 68(8):4162–4167, (2002).
- Singh, Narendra; Tang, Yuanyuan; Zhang, Zuotai; Zheng, Chunmiao; "COVID-19 waste management: Effective and successful measures in Wuhan, China," Resour Conserv Recycl. 2020, Dec.; 163: 105071.
- Spilimbergo, S., *"A study about the effect of dense CO_2 on microorganisms"*, Ph.D. thesis, University of Padova, Italy, 2002.
- STAATT, *"Report of the State and Territorial Association on Alternative Treatment Technologies (SAATT)"*, Technical assistance manual: state regulatory oversight of medical waste treatment technology, (2005).
- Watanabe, K., Kakita, Y., Kashige, N., Miake, F., Tsukiji, T., *"Effect of ionic strength on the inactivation of micro-organisms by microwave irradiation"*, Lett Appl Microbiol 31:52–56., (2000).

- WHO Health-care waste. https://www.who.int/news-room/fact-sheets/detail/health-care-waste, 2018.
- WHO, *"Management of solid healthcare waste at primary healthcare centres",* A Decision Making Guide, Geneva, (2005).
- Wu, H.S., Liu, Y.D., Zhao, G.M., Chen, X.O., Yang, X. N., Zhou, X. D., *"Succinic acid inhibited growth and pathogenicity of in vitro soil-borne fungus Fusarium oxysporum f. sp. Niveum",* Acta Agriculturae Scandinavica, Section B Soil Plant Sci 61(5):404–409., (2011).
- United Nations Environment Programme; *"Waste Management during the COVID-19"* Pandemic-from-Response-to-Recovery, 2020 , https://reliefweb.int/sites/reliefweb.int/files/resources/WMC-19.pdf.
- Zambrano-Monserrate, M.A., Ruano, M.A., Sanchez-Alcalde, L., *"Indirect effects of COVID-19 on the environment",* Sci. Total Environ. 728, 138813, 2020.

❐

The manufacturer's authorised representative in the EU is Springer Nature Customer Service Centre GmbH, Europaplatz 3, 69115 Heidelberg, Germany. If you have any concerns regarding our products, please contact ProductSafety@springernature.com

Printed and bound by CPI Group (UK) Ltd, Croydon, CR0 4YY

31/07/2026

02182889-0001